SpringerBriefs in Applied Sciences and Technology

More information about this series at http://www.springer.com/series/8884

Iraj Sadegh Amiri · Sayed Ehsan Alavi
Sevia Mahdaliza Idrus

Soliton Coding for Secured Optical Communication Link

 Springer

Iraj Sadegh Amiri
Sayed Ehsan Alavi
Sevia Mahdaliza Idrus
Lightwave Communication Research Group
  Faculty of Electrical Engineering
Universiti Teknologi Malaysia
Skudai, Johor
Malaysia

ISSN 2191-530X   ISSN 2191-5318   (electronic)
ISBN 978-981-287-160-2   ISBN 978-981-287-161-9   (eBook)
DOI 10.1007/978-981-287-161-9

Library of Congress Control Number: 2014945111

Springer Singapore Heidelberg New York Dordrecht London

Printed on acid-free paper

Springer is part of Springer Science+Business Media (www.springer.com)

Acknowledgments

The authors would like to thank the, Lightwave Communication Research Group, Faculty of Electrical Engineering, Universiti Teknologi Malaysia, 81310 UTM Skudai, Johor, Malaysia for providing the research facilities.

Contents

Abstract

Nonlinear behaviors of light such as chaos can be observed during propagation of a laser beam inside microring resonator (MRR) systems. Chaotic signals can be used to generate data of logic codes to be transmitted along the fiber optic communication. The data of logic codes generated by the single ring resonator system can be used to generate secured codes, where the decoding process of the transmitted codes can be obtained at the end of the transmission link. Thus, secured transmitting of signals can be obtained along the long distance fiber communication. We propose a system of MRRs to generate a series of logic code. Optical soliton can be used to generate entangled photon as well. The ultra-short soliton pulses are providing required communication signals to generate pair of polarization entangled photons required for quantum keys. In the frequency domain, MRRs can be used to generate optical millimetre-wave solitons with a broadband frequency of 0–100 GHz. The soliton signals can be multiplexed and modulated with the logic codes to transmit the data via a network system. The soliton carriers play critical roles to transmit the data via an optical communication link and provide many applications in secured optical communications. Therefore, transmission of data information can be performed via a communication network using soliton pulse carriers. A system known as optical multiplexer can be used to increase the channel capacity and security of the signals.

Chapter 1
Introduction of Fiber Waveguide and Soliton Signals Used to Enhance the Communication Security

J.J. Thomson in 1893 proposed first waveguide, where it was experimentally verified by O.J. Lodge in 1894. Analysis of the propagating modes was executed mathematically by Lord Rayleigh in 1897 within a hollow metal cylinder. In April 1957, the scientists tried to achieve maser-like amplification of visible light. In November of 1957, Gordon Gould, an American physicist (credited with the invention of the laser) could make an appropriate optical resonator by using two mirrors in the form of a Fabry-Perot interferometer. Unlike other designs, this new design would produce a narrow, coherent, intense beam. The gain medium could easily be optically pumped to achieve necessary population inversion. He also considered pumping of the medium by atomic-level collisions, and expected many of the potential uses of such a device [1]. Laser has a wide range of application including lidar, ladar, and communications.

Optical beam has a natural attraction to be diffracted while it is propagating in a uniform medium. The beams diffraction can be compensated by beam refraction when the refractive index is increased. Optical waveguide is a key way to present a balance between diffraction and refraction if the medium is uniform regarding to the direction of propagation. Therefore the outcome propagation of the light is controlled in the transverse direction of the waveguide, and it is described by the concept of spatially localization of the electric field in the waveguide [2].

Typically, an optical fiber consists of a transparent core which is surrounded by a transparent cladding material with a lower refractive index. Light is confined in the core by total internal reflection so it acts as a waveguide. Fibers which can be used for many propagation paths or transverse modes are called multi mode fibers (MMF) whereas those which can only support a single mode are called single mode fiber (SMF). The MMF types of fiber generally have a larger core diameter and can be used for short distance communication systems while SMF fibers are used for long distance communication greater than 1,050 m (3,440 ft) [3].

© The Author(s) 2015
I. Sadegh Amiri et al., *Soliton Coding for Secured Optical Communication Link*,
SpringerBriefs in Applied Sciences and Technology,
DOI 10.1007/978-981-287-161-9_1

Optical solitons are localized as electromagnetic waves that propagate in nonlinear media resulting from a balance between nonlinearity and linear broadening due to dispersion and/or diffraction. There are five types of nonlinear media which such as Kerr law, power law, parabolic law, dual-power law and the log law. In the presence of dispersive perturbation terms, the phenomena of optical soliton cooling are also observed. Initially soliton refer to the particle-like nature of solitary waves that remain intact even after common collisions [4]. The first observation of soliton was done by Scott Russel on the Edinburgh-Glasgow canal in 1834. He observed that a wave travelling through a canal without lost and major changes of its shape [5].

This observation appears to disagree with the nonlinear theory of Airy published in 1845, which predicted that a wave of finite amplitude cannot transmit without a change of its shape. According to his theory the wave should attenuate. The problem was solved by Joseph Boussinesq [6] in 1871. In 1876 Lord Rayleigh [7] independently, he was able to show that in a solitary wave the increase in local wave velocity associated by finite amplitude is balanced by the decrease associated with dispersion. In 1895, Korteweg de Vries [8] developed a model which can explain the unidirectional propagation of the waves of long wavelength in water with relatively shallow depth. This equation now is known as KdV. However the properties of solitons are not clearly understood until several mathematical models were introduced. The inverse scattering method was developed in the 1960s and it was able to explain the properties of soliton. The mathematical solution of soliton as KdV was found by Zabusky and Kruskal in 1964 [9].

Several model equations of soliton phenomena can be presented along with Soliton solutions. In 1989, Drazin and Johnson described solitons as solutions of nonlinear differential equations which represent waves of permanent form, localized, (so that they decay or approach a constant at infinity), strongly interactive with other solitons, where they emerge from the collision unchanged apart from a phase shift. Many exactly solvable models have soliton solutions, including the Korteweg-de Vries equation, the nonlinear Schrödinger equation, the coupled nonlinear Schrödinger equation, and the sine-Gordon equation. The soliton solutions are typically obtained by means of the inverse scattering transform. The mathematical theory of these equations is a broad and very active field of mathematical research [10, 11].

In 1973 it was discovered that optical fibers can support dark solitons when the group-velocity dispersion (GVD) is "normal". Hasegawa and Tappert could solve and explain the non-linear Shrödinger (NLS) equation and the theory of optical soliton [12]. The first generation of spatial solitons was reported in 1974 by Ashkin and Bjorkholm [13] in a cell filled with sodium vapor. Only a decade later, Mollenauer performed the first experiment of soliton propagation in optical fibers due to the lack of adapting low loss fibers at that time [14]. Temporal dark solitons became very interesting during the 1980s [15]. During the decade of the 1990s, many other kinds of optical solitons such as spatiotemporal, Bragg, vortex, vector and quadratic solitons were discovered.

In the most recent overview of experimental observations of spatial optical solitons, some materials display large optical non-linearities when their properties are customized by the light propagation [16]. Particularly, if the non-linearity causes a change of the refractive index of the medium in which the beam can become self-trapped and propagates unchanged exclusive of any external waveguiding structure. These types of stationary self-guided beams are known as spatial optical solitons [17].

The NLS equation is much newer than the KdV equation [18]. The NLS equation without the time-derivative term first appeared in 1950 within the context of superconductivity [19], where in the 1960s, the stationary version was used to describe the self-focusing of light in nonlinear media [20]. A time-dependent version first appeared in 1961, where it was used to describe Bose condensates in solid state physics [21]. Nevertheless, Zakharov [22], in 1967, for the first time used the time-dependent NLS equation to describe the evolution of optical wave packets in nonlinear dielectric media, where the one-dimensional, time-dependent version first appeared in 1967 [23].

In 1970s, Hasegawa primed to realize that the NLS equation was appropriate for the calculation of pulse propagation in optical fibers, and that they should therefore support solitons. In a seminal work published in 1973 [24], he and co-author Frederick Tappert showed how the NLS equation applied to single-mode fibers, derived the essential properties of the corresponding solitons. In supporting numerical simulation they showed that the solitons were stable and robust. It is noteworthy that at the time, fibers having low loss in the region of anomalous dispersion ($\lambda > 1,300$ nm) did not exist. Hasagewa and Tappert followed up almost immediately with another paper [25] describing dark solitons, i.e., sech-shaped holes in a CW background, which could exist in the presence of normal dispersion. For a number of practical reasons, however, the dark solitons have never been used for transmission.

The first experimental observation of soliton [26] was occurred by using microscope objectives. It was done when the mode-locked color center laser's output was coupled into the fiber, and the fiber's output into an autocorrelator.

The shape of fiber is adjustable, thus the ring resonators can be made and used to resonate selective wavelength or can be used as filters. Microring resonators shaped from nanoscale photonic waveguides. Ring resonators are employed to generate signals used for optical communication applications, where they can be integrated in a single system. Optical microring resonators recently are interesting subject in the area of integrated optics because of their unique aspects such as compactness, low cost, tunability and easy integration on a chip with other photonic devices, having a variety of applications such as optical filter, optical switch, optical modulator, optical delay line, dispersion compensator, optical sensor and etc. [27].

Since ring resonators are used to support the travelling of wave resonant modes, a single ring may be applied to completely extract a particular wavelength from a signal bus. Therefore they are ideal candidates for very large-scale integrated (VLSI) photonic circuits, since they provide a wide range of optical signal processing functions while being ultra compact [28]. Microring resonators have good advantages when they are used as a filter system [29]. The type of semiconductor microring is used widely to enhance the nonlinear optical effects which are proposed and investigated [30].

Ring resonators are not used only in optical networks, but they have recently been presented to be used as sensors, filters and biosensors. There are many research works on the fabrication and characterization of integrated ring resonators in a variety of material systems. The first thesis on integrated ring and disk resonator filters explains the fabrication and characterization of devices with diameters smaller than 10 μm is investigated in the material system AlGaAs/GaAs. First theoretical works were written in 1998, where the finite difference time domain (FDTD) analysis of the ring and disk resonators was presented. In 2000 Absil wrote a thesis based, on the material system AlGaAs/GaAs, where a vertical and lateral multiple coupled ring resonator configuration could be fabricated and characterized. Most of the research works on the ring resonator in the micro and nano size scale have been done since 2000. The security and high capacity of optical communication network is the major concern in the field of nano photonics.

A realistic perfect communication transmission security can be achieved by the technique known as quantum key distribution (QKD). Several research methodologies have been reported showing promising achievements of QKD in various systems. Suchat et al. [31] could propose the use of QKD via optical wireless link. In this study the secure information for instance, telephone conversation could be secured which is used in telephone networks.

Yupapin et al. presented the idea of quantum-chaotic encoding when optical soliton pulse travels in a fiber optic ring resonator. They could find that the random bits of information can be generated, where it is significant enough to be used as security codes for network communication applications. This introduces the double security which is known as quantum-chaotic. In this progress the pumping input power can be operated to generate the chaotic behavior of the propagating light within the fiber ring resonator, where the nonlinear condition of Kerr effect is induced into the system. The output signals of the system can be used to create two different codes. One is the codes of quantum bits for instance, while the other is introduced as the chaotic signal. The advantage of the system is that it has high potential of using for communication security. If this technique is used to incorporate with the optical communication link, the quantum-chaotic codes can be generated. It means that the high capacity of secured information can be realized [32].

Mitatha et al. showed that practically, the chaotic codes can be formed by quantizing the chaotic signals. It is shown that the chaotic signals can be switched (On/Off) via the band pass or band stop filters to the specific users. Therefore, the high-capacity and secured communication data can be performed in the optical network systems. Simulated results have shown that there are two schemes of the chaotic codes, which can be used to generate 100 logical codes. The random method can be used when the random input is used in order to control the optical chaotic signals and threshold powers. Here the ultra-fast switching time, in the range of 10^{-15} s (fs), was obtained, which is fast compared with the switching time from electronic scheme.

As advantage of such a system, the optical signals can be randomly chaotic encoded and switched for the specified users. Suitable techniques for encoding the chaotic signals is known as synchronized technique. In this technique, the required message can be successfully decoded by deducting the chaotic oscillation. This can be implemented by the receiver in the transmitted signal using the least-squares method. It was demonstrated that the chaotic signal is reasonably encoded by using the waveforms of the transmitter and chaotic signals of the receiver output. Thus, conducting of a secure transmission of a message and logical coding can be done using chaotic quantizing and coding [33].

In 2010, Thongmee and Yupapin [34] have shown that the secured optical communication can be done when the chaotic signals are generated in a nonlinear micro-ring resonator system. The variable parameters of such as systems are used to attain desired results. Mitatha et al. [35] have proposed the design of secured packet switching using nonlinear behaviors of light. This proposed system can be used for high-capacity and security switching in optical communication network. Some other related research works have been reported in various forms of applications. However, a more applicable system for network security is needed, which can satisfy both secure and high capacity.

Pornsuwancharoen proposed a novel system of a simultaneous generation of continuous variable quantum key distribution (QKD) and quantum dense coding (QDC) by means of an optical memory array. They have reported that light pulses can be stored within a nano-waveguide, which is offered for quantum key generation source. Moreover, the continuous variable QKD can also be catered via the system. QKD has been recognized as the top secret key for information security purposes. This method is the best way to keep the information in secret used for communication and network security. Many research works have been proposed and investigated in both theoretical and experimental works. This research shows promising results that can be used to make the realistic security [36, 37].

A new design of a security method using the nonlinear behaviors of the dark and bright soliton in the form of temporal pulses is proposed by W. Siririth et al. The proposed system consists of micro-ring resonators which can be used for signal security application. Another aspect of this study is that by using the appropriate ring parameters, the soliton conversion can be performed. In this study different response time of the temporal soliton was found. As a result, the response times of 169 ns and 84 ns were obtained for the temporal type of the dark and bright solitons, respectively. Response time can be used to form the security key in communication networks.

Thus, in this study the dark soliton pulse propagating within the optical media can be converted into a bright soliton pulse when the system of ring resonator is incorporated with an add/drop multiplexer system. By using the reasonable dark soliton input power, a bright soliton could be obtained used to provide the common soliton for long-distance communication. As a conclusion, it was investigated and proved that, the use of dark soliton to form the signal security or communication security is possible [38]. Therefore conversion of the dark soliton leads to generate the common type of the soliton as a bright soliton which is used for long-distance communication link.

K. Sarapat et al. proposed a new idea of quantum soliton pulse generation using MRRs. The used system was made of a series of micro ring resonators connected to numbers of beam splitters. Firstly, the chaotic optical soliton pulses were generated and secondly, the specific second harmonic pulse was selected using appropriate parameter of the rings. The advantage of such a system is that the quantum repeater unit can be unnecessary for long distance quantum communication link, whereas the using the system for multi-entangled photon sources is available. Here the high-power second harmonic soliton pulses can be performed with the strong entangled photon visibility. Clear and strong entangled photons are generated and capable for multi- and long distance links [39].

Dark soliton has a great attention in optical communication, where the security of the transmission data can be performed. Charoenmee et al. found that a dark soliton pulse can be localized within a nonlinear nano-waveguide. The used system consists of nonlinear micro and nano ring resonators, whereas the dark soliton could be input into the system and trapped within the nano-waveguide. In this work a dark soliton pulse is input into a ring resonator being chopped to smaller pulses. Required pulse is filtered and amplified, which can be controlled and localized within the nano-waveguide.

Polthep Srimuk et al. proposed a new design of a security camera system in order to increase the channel capacity and security of the output signals. This system uses the dense wavelength division multiplexing wavelength enhanced. Increasing in number of channels can be obtained by increasing in wavelength density, while the security is introduced by the specific wavelength filtering. The main advantage of the study is that the proposed system can be implemented and incorporated with the existed communication link in either wire/wireless system, in which the human privacy can be provided. This system allows for the increasing channel capacity by using optical technique DWDM.

The security effects can be provided when specific wavelengths are used and needed to be filtered in order to retrieve the required signals. As a result, the multi-wavelength bands can be generated by using a Gaussian pulse propagating within the MRR system. The obtained results are available for the extended DWDM when the wavelength center of 1,300 and 1,500 nm, are used for the existed public networks. Therefore, the increasing in wavelength capacity can be used to increase the CCTV channels in which the human privacy can be maintained when the specific wavelength filter is used. Interesting results of signals with FWHM and FSR of 30 pm and 600 pm respectively could be generated and achieved [40].

Communication security needs the security and privacy requirements due to the large demand of world networks. Therefore the security schemes, such as quantum and optical techniques, have widely used secured communications. An interesting security technique is presented by Juleang et al. to generate public key suppression, where the dark-bright soliton conversion control within a PANDA system is used. The public key suppression and public key recovery can be used in a highly secure communication system and has potential applications in optical cryptography [41].

A new technique of generation ghost-signal by microring resonator is investigated by Dunmeekeaw et al. The proposed system is made of a series of micro ring resonators connecting to an add/drop filter system, where signal security is generated for optical communication system. The input pulse of Gaussian beam with central wavelength of 1.3 μm can be input into the system, where chaotic signals can be generated and multiplexed into the optical communication link. Here, results were generated to identify two similar "signal" and "ghost" signals, which are observed in a different time frame. In this application, communication security can be performed when the required information is multiplexed using the chaotic signal, signal and ghost signals [42].

Juleang et al. proposed a novel design of optical network system, made by micro ring resonators used for optical communication security and receiver identification. The advantage of this research is that data signals can be secured by using both optical encryption key and optical address of the receiver before transmission. On the receiver part, a system of the decryption key is needed to recover the transmitted information. For the aim of encryption, decryption and identification dark-bright soliton pulses are propagating within PANDA ring resonator, optical add/drop filter and Mach–Zehnder Interferometer. Simulation results obtained have shown that the secured optical network can be achieved optically and realized, where the dynamic behavior of dark-bright solution propagation and collision within the ring system can be used to generate chaotic signals. The chaotic signals can be used to generate optical encryption and decryption keys using suitable add/drop system. The MZIs is used to secure (encrypt/decrypt) the optical communication [43].

Micro and nanoscale devices have been used widely in information technology such as telephone handsets. One of the important parts of this device is an antenna. Using a small antenna with good performance is necessary, where to date, nano-antenna has become an interesting field in many applications such as biology and medicine. A novel nano-antenna system design is presented by Thammawongsa et al. in which photonic spins in a PANDA ring resonator are employed. These spins are generated using soliton pulse within a PANDA system. The magnetic field is introduced by using an aluminum plate coupling to the microring resonator, in which the spin-up and spin-down states are induced, where finally, the photonic dipoles are formed. The advantage of the proposed system is that powerful simple and compact nano-antenna can be fabricated. In addition, optical dipole can be used for further research such as dynamic dipole, dynamic torque, nano-motor, spin communicated and spin cryptography, etc. [44].

The use of data and information in optical communication is growing day by day. Therefore, the security of data is a major concern, where there are a lot of techniques which can be used to protect the secret data or information. Up to date, a quantum technique is recommended to provide such a requirement. A new concept of quantum cryptography using dark-bright soliton conversion behaviors within a nonlinear ring resonator (PANDA ring resonator) is presented by Tunsiri et al. In this research orthogonal soliton is established among the soliton conversion. The advantage of this research is that long distance quantum communication and high capacity quantum communication can be performed using the powerful entangle soliton. In application, the high capacity quantum communication is variable by using the multi variable entangled solitons [45].

Chaotic signals have some properties such as broadband, orthogonality and complexity aspects, which prompt researches in the areas of nonlinear science, communication technology and signal processing [46, 47]. The concern in chaotic communications was due to the foreseen good properties of the chaotic signals in the fields of security systems or broadband multiple access systems. The possibility of employing chaotic signals to carry information was first studied in 1993. Encoding is the process of adding the correct transitions to the message signal in relation to the data [48] that is to be sent over the communication system. Fiber optic sensors and micro structured fibers hold great promise for integration of multiple sensing channels. Nonlinear behavior of light inside a microring resonator (MRR) takes place when a strong pulse of light is inserted into the ring system [49, 50]. Chaotic controls have been used in a great number of optical, engineering and biological designed systems [51–53].

Encoding is used in binary search algorithms to determine where the collided bit is. A method of encoding data into a chain reaction code includes generating a set of input symbols from input data. Grover has provided fundamental information-theoretic bounds on the required circuit wiring complexity and power consumption for encoding and decoding of error-correcting codes. In encoding method, the negative edge of signal means data-1, positive edge of signal means data-0. By decoding method, signals will not be changed, if a collision occurs. So, the reader can easily find the signals which are not changed for identifying the collision. That is, the reader can determine easily where the collision bits are. In this work, generation of chaotic signals in a single ring resonator is presented. Thus the single ring resonator system can be used to generate demand logic codes where the technique of encoding and decoding of transmitting information via optical logic signals can be obtained [54, 55]. Transmitting of signals can be secured throughout the propagation along optical fiber communication in which original and initial signals are recovered using the encoding-decoding technique [56].

Photon switching is concerning field of optical communication, where it employs quantum cryptography in a mobile telephone network, which is described by Amiri et al. [57, 58]. Quantum key distribution supplies the perfect communication security. Hence quantum cryptography can be performed through an optical-wireless link. Research works have shown that techniques of continuous variable quantum cryptography are aimed and applied on the MRRs [59, 60]. Entangled photon pairs are an important resource in quantum optics, and are essential for quantum information applications such as quantum key distribution and controlled quantum logic operations. Furthermore, control over the pair generation time is essential for scaling many quantum information schemes beyond a few gates. New quantum key distribution protocol points that data can be encoded on continuous variables of a single photon. In order to give rise a range of light throughout a wide range, an optical soliton signal is suggested as an improved laser pulse used to make chaotic filter characteristics when propagate inside MRRs [61–63]. The capacity of the system can be increased when the chaotic packet switching is employed. We propose a system that purposes localized soliton pulses to figure the high capacity and security communication. It is used to trap optical solitons to generate entangled photon pair. Furthermore, the continuous variable quantum codes can be generated using the polarizer and beam splitter systems.

Optical communication is an interesting area in photonics for two decades. It is very attractive especially when it uses quantum cryptography in a network system where it was reported by Amiri et al. [64, 65]. Quantum keys can form requires information which provides the perfect communication security. Amiri et al. showed that quantum security could be performed via the optical-wired and wireless link [66]. Some research works have shown that some techniques of quantum cryptography are proposed, where the systems of MRR are still complicated. Amiri et al. [67] proposed a new quantum key distribution rule in which carrier information is encoded on continuous variables of a single photon. In such a way, Alice randomly encodes information on either the central frequency of a narrow band single-photon pulse or the time delay of a broadband single-photon pulse. Liu and Goan studied the entanglement evolution under the influence of non-Markovian thermal environments. The continuous variable systems could be two modes of electromagnetic fields or two nano-mechanical oscillators in the quantum domain, where there is no process that can be performed within a single system.

To generate a spectrum of light over a broad range, an optical soliton pulse is recommended as a powerful laser pulse that can be used to generate chaotic filter characteristics when propagating within MRRs [68–70]. Therefore, the capacity of the transmission data can be secured and increased when the chaotic packet switching is employed. In this research, we simulate localized spatial and temporal soliton pulses to form the high capacity and security communication [71, 72]. The MRR system is used to trap optical solitons in order to generate entangled photon pair required for quantum keys. Here, generation of the localized ultra-short soliton pulses for continuous variable application is demonstrated [73–75]. The system of quantum key generation can be implemented within the wireless networks. Thus, the links can be set up using the optical soliton, generated by the technique called chaotic filtering scheme in which required signals can be selected and used [76, 77]. The device parameters are simulated according to the practical device parameters, where the results obtained have shown that the entangled photon pair can be performed within the MRR device.

Orthogonal frequency division multiplexing (OFDM) is a combination of modulation and multiplexing. Modulation refers to the process of changing the carrier phase, frequency, amplitude, or their combination with a modulated signal that typically contains information to be transmitted. However, the aim of multiplexing is to share a bandwidth. Single-carrier modulation is a technology that modulates information onto only one carrier. The main problem of this technology is satisfying the need for high bandwidth in fixed spectrum limits of one single-carrier. High data rate in one carrier causes a high symbol rate. As the duration of one symbol or bit becomes smaller, the system becomes more susceptible to loss of information from impulse noise, signal reflections, and other impairments. These impairments can impede the ability to recover information sent. In addition, as the bandwidth used by a single-carrier system increases, the susceptibility to interference from other continuous signal sources becomes greater. This type of interference is commonly labelled as a carrier wave (CW) or frequency interference [78–80].

In OFDM instead of transmitting symbols over the communication channel, the channel is split into many sub-channels, and the digital symbols are transmitted in parallel over these sub-channels. OFDM provides both a high data rate and symbol duration using frequency division multiplexing (FDM) over multiple subcarriers within one channel. Furthermore, this technique exploits subcarriers that are orthogonal to each other. In follow section, we will describe orthogonality in more detail. In a traditional parallel data-stream, bandwidth can be divided into sub-channels, where a fixed channel spacing between them is needed to eliminate inter-carrier interference (ICI). This technique results in exiting spectrum waste. In contrast, OFDM uses overlapping subcarriers without causing ICI. To achieve this goal, in OFDM, subcarriers are orthogonal to each other.

Besides high spectral efficiency, OFDM has high tolerance to multi-path interference, channel dispersion and frequency-selective fading. Moreover, because of dynamic bandwidth allocation and adaptive bit rate functionalities OFDM system has prominent flexibility. OFDM implementation is based on high-speed digital signal processing (DSP) and by development of high speed DSPs at several Gb/s and higher, strong interest in applying the OFDM technique in optical communication field has been stimulated. Optical OFDM (O-OFDM) has superior robustness to fiber chromatic dispersion and polarization mode dispersion (PMD). In addition high spectral efficiency using higher-order modulation which enables dynamic data rate adaptation is feasible in O-OFDM.

Based on system complexity and cost, for applications in different segments of optical networks O-OFDM can be implemented using intensity modulation (IM) with direct detection (DD-O-OFDM) or coherent OFDM (CO-O-OFDM) [81]. The former has a simple receiver, but some optical frequencies must be unused if unwanted mixing products are not to cause interference. This is usually achieved by inserting a guard band between the optical carrier and the OFDM subcarriers. This reduces spectral efficiency and also requires more transmitted optical power, as some power is required for the transmitted carrier. The later requires a laser at the receiver to generate the carrier locally, and is more sensitive to phase noise but with better spectral efficiency.

In the last few years, power system dynamics have been studied from the nonlinear dynamics point of view, using chaotic theory. Nonlinear light behaviour inside an MRR occurs when a strong pulse of light is inputted into the ring system; this is used for many applications in signal processing and communication [82–84]. The properties of a ring system can be modified via various control methods. Ring resonators can be used as filter devices where trapping of optical frequency or wavelength can be obtained using suitable system parameters [85, 86]. The panda ring resonator, which consists of a centred ring resonator connected to two smaller ring resonators on the right and left sides, is used in many applications in optical communication and signal processing. This system can be used to generate optical soliton pulses of millimetre wave range and GHz frequency, thus providing required signals used in wired/wireless optical communication. Therefore, MRRs can be used to generate a broad band of optical solitons, applicable in many areas of optical communication networks, such as wireless cable systems and indoor–outdoor communication [87–90].

A network system can be designed to provide transmission of secret data with the highly efficient transmission of soliton signals based on the OFDM application. In this work, the optical soliton in a nonlinear fibre MRR system is analysed in order to generate a high frequency band of pulses to be multiplexed with generated logic codes from chaotic signals, using an OFDM technique. Control of the process can be achieved by controlling the parameters of the system, such as round trip, input power, coupling loss, coupling coefficient, the ring's radius, and linear/nonlinear refractive indices.

Digital multiplexing such as Time Division Multiplexing (TDM) is a type of network system in which two or more bit signals can be transferred simultaneously as sub-channels in one communication channel [91–93]. It can be further extended into the time division multiple access (TDMA) system, where several stations connected to the same physical medium, for example sharing the same frequency channel, can communicate.

Therefore, a TDMA system can be realized as a channel access method for shared medium networks, where the users receive information with different time slots. This allows multiple stations to share the same transmission medium while using only a part of its channel capacity [94]. TDMA can be used in digital mobile communications and satellite systems. Thus, in the TDMA system, instead of having one transmitter connected to one receiver, there are multiple transmitters, where high-secured signals of quantum codes along the users can be transmitted. Secured communication is increasing widely and rapidly every year. The security technique known as quantum cryptography has been widely used and investigated in many applications. Hence, the internet security becomes an important function which is required to be included in the modern internet service. So far, a quantum technique is recommended to provide such a requirement. Potential of using the optical tweezers and quantum codes, especially, for the hybrid quantum communication in the network system is expressed [95–97]. Quantum codes can be performed and generated via optical tweezers in the form of potential wells with appropriate soliton input power and MRR parameters. A new technique for Quantum Key Distribution (QKD) was presented by Amiri et al. that can be used to make the communication transmission security and implemented by a small device such as mobile telephone hand set. This technique uses the Kerr nonlinear type of light in the MRR. Mitatha et al. have proposed the design of secured packet switching used nonlinear behaviors of light in MRR which can be made for high-capacity and security switching [98–100].

To date quantum code is the only form of information that can provide the perfect communication security. Dark-Gaussian soliton controls within a semiconductor add/drop multiplexer has numerous applications. Optical tweezers technique also becomes a powerful tool for manipulation of micrometer-sized particles in three spatial dimensions and has led to widespread applications in biology, and in physical sciences [101–103]. The output is formed when the high optical field is configured as an optical tweezers or potential wells. Optical tweezers in the forms of valleys (potential wells) are kept in the stable form within the add/drop filter. Several emerging technologies, such as integrated all optical signal processes and all-optical quantum information processing, requires strong and rapid interactions between two distinct optical signals. In this book, we have used a nonlinear MRR to form the high secured correlated quantum codes. Here the novel system of dynamic optical tweezers (potential wells) generation, using dark soliton pulses propagating within an add/drop multiplexer is presented. Due to low power of the potential wells, the output signals from the proposed system can be highly secured during propagation along the communication network. Therefore, signals in the form of digital codes can be detected by different users. Here, different time switching of the signals could be obtained using different fiber length, connecting the digital signal transmitter system to the users [104–106].

References

1. Hammond B et al (2002) Integrated wavelength locker for tunable laser applications. IEEE
2. Snyder AW, Love JD (1983) Optical waveguide theory, vol 190. Springer
3. Hecht J (2010) Short history of laser development. Opt Eng 49:091002
4. Abdullaev F, Garnier J (2005) Optical solitons in random media. Prog Opt 48:35–106
5. Chiao RY, Garmire E, Townes C (1964) Self-trapping of optical beams. Phys Rev Lett 13(15):479
6. Narahara K, Nakagawa S (2010) Nonlinear traveling-wave field effect transistors for amplification of short electrical pulses. IEICE Electron Express 7(16):1188–1194
7. Sander J, Hutter K (1991) On the development of the theory of the solitary wave. A historical essay. Acta Mech 86(1):111–152
8. Israwi S (2010) Variable depth KdV equations and generalizations to more nonlinear regimes. ESAIM Math Model Numer Anal 44(02):347–370
9. El G, Grimshaw R, Smyth N (2009) Transcritical shallow-water flow past topography: finite-amplitude theory. J Fluid Mech 640:187–214
10. Hasegawa A (2002) Optical solitons in fibers for communication systems. Opt Photonics News 13(2):33–37
11. Maimistov AI (2010) Solitons in nonlinear optics. Quantum Electron 40:756
12. Zhang XF, He WQ, Zhang P (2011) Controllable optical solitons in optical fiber system with distributed coefficients. Commun Theor Phys 55:681
13. Wise FW (2001) Spatiotemporal solitons in quadratic nonlinear media. Pramana 57(5):1129–1138
14. Mollenauer L, Smith K (1988) Demonstration of soliton transmission over more than 4,000 km in fiber with loss periodically compensated by Raman gain. Opt Lett 13(8):675–677
15. Stratmann M, Mitschke F (2005) Chains of temporal dark solitons in dispersion-managed fiber. Phys Rev E Stat Nonlin Soft Matter Phys 72(6 Pt 2):066616
16. Fischer R et al (2006) Observation of spatial shift in interaction of dark nonlocal solitons. IEEE
17. Stegeman GI, Segev M (1999) Optical spatial solitons and their interactions: universality and diversity. Science 286(5444):1518–1523
18. Schneider G (2011) Justification of the NLS approximation for the KdV equation using the Miura transformation. Adv Math Phys
19. Ginzburg VL (1955) On the theory of superconductivity. Il Nuovo Cimento (1955–1965) 2(6):1234–1250
20. Carter JD, Contreras CC (2008) Stability of plane-wave solutions of a dissipative generalization of the nonlinear Schrödinger equation. Physica D 237(24):3292–3296
21. Gross EP (1963) Hydrodynamics of a superfluid condensate. J Math Phys 4:195
22. Zakharov V (1967) On instability of light self-focusing. Zh Eksp Teor Fiz 53:1743–1745

23. Benney D, Newell A (1967) The propagation of nonlinear wave envelopes. J Math Phys 46(2):133–139
24. Hasegawa A, Tappert F (1973) Transmission of stationary nonlinear optical pulses in dispersive dielectric fibers. I. Anomalous dispersion. Appl Phys Lett 23(3):142–144
25. Hasegawa A, Tappert F (1973) Transmission of stationary nonlinear optical pulses in dispersive dielectric fibers. II. Normal dispersion. Appl Phys Lett 23:171
26. Mollenauer LF, Stolen RH, Gordon JP (1980) Experimental observation of picosecond pulse narrowing and solitons in optical fibers. Phys Rev Lett 45(13):1095–1098
27. Little BE et al (1997) Microring resonator channel dropping filters. J Lightwave Technol 15(6):998–1005
28. Daldosso N, Pavesi L (2009) Nanosilicon photonics. Laser Photonics Rev 3(6):508–534
29. Liang D et al (2011) Hybrid silicon ring lasers
30. Absil P et al (2000) Wavelength conversion in GaAs micro-ring resonators. Opt Lett 25(8):554–556
31. Suchat S, Khunnam W, Yupapin PP (2007) Quantum key distribution via an optical wireless communication link for telephone networks. Opt Eng 46:100502
32. Yupapin PP, Chunpang P (2009) A quantum-chaotic encoding system using an erbium-doped fiber amplifier in a fiber ring resonator. Optik-Int J Light Electron Opt 120(18):976–979
33. Kues M et al (2009) Nonlinear dynamics of femtosecond supercontinuum generation with feedback. Opt Express 17(18):15827–15841
34. Thongmee S, Yupapin P (2010) Chaotic soliton switching generation using a nonlinear micro ring resonator for secure packet switching use. Optik-Int J Light Electron Opt 121(3):281–285
35. Mitatha S et al (2010) High-capacity and security packet switching using the nonlinear effects in micro ring resonators. Optik-Int J Light Electron Opt 121(2):159–167
36. Mattle K et al (1996) Dense coding in experimental quantum communication. Phys Rev Lett 76(25):4656
37. Pongwongtragull P, Mitatha S, Yupapin P (2010) A simultaneous generation of QKD and QDC via optical memory array for distributed network security. Optik-Int J Light Electron Opt 121(23):2137–2139
38. Siririth W et al (2010) A novel temporal dark-bright solitons conversion system via an add/drop filter for signal security use. Optik-Int J Light Electron Opt 121(21):1955–1958
39. Sarapat K, Pornsuwancharoen N, Yupapin P (2010) Polarized soliton pulses generation using nonlinear micro ring resonators for multi-and long distance links. Optik-Int J Light Electron Opt 121(6):553–558
40. Srimuk P, Mitatha S, Yupapin PP (2010) Novel CCTV security camera system using DWDM wavelength enhancement. Procedia-Soc Behav Sci 2(1):79–83
41. Juleang P et al (2011) Public key suppression and recovery using a PANDA ring resonator for high security communication. Opt Eng 50:035002
42. Dunmeekeaw U et al (2012) A new technique generation ghost-signal by microring resonator for 1.3 μm security communication. Procedia Eng 32:516–521
43. Juleang P et al (2012) Highly secured optical communication by optical key and identification address. Optik-Int J Light Electron Opt
44. Thammawongsa N et al (2012) Novel nano-antenna system design using photonic spin in a panda ring resonator. Prog Electromagnet Res 31:75–87
45. Tunsiri S et al (2012) Optical-quantum security using dark-bright soliton conversion in a ring resonator system. Procedia Eng 32:475–481
46. Amiri IS et al (2014) Chaotic carrier signal generation and quantum transmission along fiber optics communication using integrated ring resonators. Quantum Matter
47. Alavi SE et al (2013) Chaotic signal generation and trapping using an optical transmission link. Life Sci J 10(9s):186–192
48. Amiri IS et al (2012) Generation of quantum photon information using extremely narrow optical tweezers for computer network communication. GSTF J Comput (joc) 2(1)
49. Teeka C et al (2011) ASK-to-PSK generation based on nonlinear microring resonators coupled to One MZI Arm. AIP Conf Proc 1341(1):221–223

50. Ali J et al (2010) Dark and bright soliton trapping using NMRR. ICEM2010: Legend Hotel, Kuala Lumpur, Malaysia
51. Sanati P et al. (2013) Femtosecond pulse generation using microring resonators for eye nano surgery. Nanosci Nanotechnol Lett 6
52. Amiri IS, Ali J (2013) Optical buffer application used for tissue surgery using direct interaction of nano optical tweezers with nano cells. Quantum Matter 2(6):484–488
53. Shahidinejad A. et al (2012) Network system engineering by controlling the chaotic signals using silicon micro ring resonator. In: Computer and communication engineering (ICCCE) Conference. IEEE Explore, Malaysia
54. Amiri IS, Ali J (2013) Data signal processing via a Manchester coding-decoding method using chaotic signals generated by a PANDA ring resonator. Chin Opt Lett 11(4):041901–041904
55. Gifany D et al (2013) Logic codes generation and transmission using an encoding-decoding system. Int J Adv Eng Technol (IJAET) 5(2):37–45
56. Amiri IS et al (2012) Digital binary codes transmission via TDMA networks communication system using dark and bright optical soliton. GSTF J Comput (JoC) 2(1):12
57. Kouhnavard M et al (2010) QKD via a quantum wavelength router using spatial soliton. AIP Conf Proc 1347:210–216
58. Ali J et al (2010) Quantum internet via a quantum processor. In: International conference on photonics 2010 (ICP 2010), Langkawi, Malaysia
59. Shahidinejad A et al (2014) Quantum cryptography coding system for optical wireless communication. J Optoelectron Adv Mater
60. Ali J et al (2010) Quantum signal processing via an optical potential well. In: International conference on enabling science and technology, Nanotech Malaysia, Kuala Lumpur, Malaysia
61. Amiri IS, Ali J (2014) Generating highly dark-bright solitons by gaussian beam propagation in a PANDA ring resonator. J Comput Theor Nanosci (CTN) 11(4):1–8
62. Ali J et al (2010) Generation of tunable dynamic tweezers using dark-bright collision. In: International conference, ICAMN, Prince Hotel, Kuala Lumpur, Malaysia
63. Afroozeh A, Amiri IS, Zeinalinezhad A (2014) Micro ring resonators and applications. In: Jian A (ed) LAP LAMBERT Academic Publishing, Saarbrücken, Germany
64. Amiri IS et al (2012) Quantum entanglement using multi dark soliton correlation for multivariable quantum router. In: Moran AM (ed) Quantum entanglement. Nova Science Publisher, New York, pp 111–122
65. Amiri IS et al (2012) secured transportation of quantum codes using integrated PANDA-Add/drop and TDMA systems. Int J Eng Res Technol (IJERT) 1(5)
66. Amiri IS et al (2014) Quantum transmission of optical Tweezers via fiber optic using half-panda system. Life Sci J 10(12s):391–400
67. Amiri IS et al (2013) Optical quantum transmitter with finesse of 30 at 800-nm central wavelength using microring resonators. Opt Quant Electron 45(10):1095–1105
68. Afroozeh A et al (2010) Optical memory time using multi bright soliton. In: International conference on experimental mechanics (ICEM), Kuala Lumpur, Malaysia
69. Ali J et al (2010) Proposed molecule transporter system for Qubits generation. In: International conference on enabling science and technology, Nanotech Malaysia, Malaysia
70. Shojaei AA, Amiri IS (2011) Soliton for radio wave generation. In: International conference for nanomaterials synthesis and characterization (INSC), Kuala Lumpur, Malaysi
71. Ali J et al (2010) Trapping spatial and temporal soliton system for entangled photon encoding. In: International conference on enabling science and technology, Nanotech Malaysia, Kuala Lumpur, Malaysia
72. Suwanpayak N et al (2010) Tunable and storage potential wells using microring resonator system for bio-cell trapping and delivery. AIP Conf Proc 1341:289–291
73. Amiri IS, Gifany D, Ali J (2013) Ultra-short multi soliton generation for application in long distance communication. J Basic Appl Sci Res (JBASR) 3(3):442–451
74. Amiri IS et al (2012) Ultra-short of pico and femtosecond soliton laser pulse using microring resonator for cancer cells treatment. Quantum Matter 1(2):159–165

75. Amiri IS et al (2011) Up and down link of soliton for network communication. In: National science postgraduate conference (NSPC), Universiti Teknologi Malaysia
76. Ali J et al (2010) MRR quantum dense coding. In: International conference on enabling science and technology, Nanotech Malaysia, KLCC, Kuala Lumpur, Malaysia
77. Amiri IS, Ali J (2013) Single and multi optical soliton light trapping and switching using microring resonator. Quantum Matter 2(2):116–121
78. Alavi SE et al (2014) All optical OFDM generation for IEEE802. 11a based on soliton carriers using microring resonators. IEEE Photonics J 6(1)
79. Amiri IS, Nikoukar A, Alavi SE (2014) Soliton and radio over fiber (RoF) applications. In: Jian A (ed) LAP LAMBERT Academic Publishing, Saarbrücken, Germany
80. Amiri IS et al (2013) Transmission of data with OFDM technique for communication networks using GHz frequency band soliton carrier. IET Commun
81. Armstrong J (2009) OFDM for optical communications. J Lightwave Technol 27(3):189–204
82. Jalil MA et al (2011) All-optical logic XOR/XNOR gate operation using microring and nanoring resonators. Global J Phys Express 1(1):15–22
83. Amiri IS et al (2010) Controlling center wavelength and free spectrum range by MRR Radii. In: Faculty of science postgraduate conference (FSPGC), Universiti Teknologi Malaysia
84. Amiri IS, Ali J (2013) Controlling nonlinear behavior of a SMRR for network system engineering. Int J Eng Res Technol (IJERT) 2(2)
85. Afroozeh A et al (2010) Dark and bright soliton trapping using NMRR. In: International conference on experimental mechanics (ICEM), Kuala Lumpur, Malaysia
86. Ali J et al (2010) DWDM enhancement in micro and nano waveguide. In: AMN-APLOC international conference, Wuhan, China
87. Amiri IS (2012) Dark-bright solitons conversion system for secured and long distance optical communication. IOSR J Appl Phys (IOSR-JAP) 2(1):43–48
88. Amiri IS, Nikoukar A, Ali J (2013) GHz frequency band soliton generation using integrated ring resonator for WiMAX optical communication. Opt Quant Electron
89. Kouhnavard M et al (2010) Soliton signals and the effect of coupling coefficient in MRR systems. In: Faculty of science postgraduate conference (FSPGC). Universiti Teknologi Malaysia
90. Shahidinejad A et al (2014) Solitonic pulse generation for inter-satellite optical wireless communication. Quantum Matter 3(2):150–154
91. Shojaei AA, Amiri IS (2011) DSA for secured optical communication. In: International conference for nanomaterials synthesis and characterization (INSC), Kuala Lumpur, Malaysia
92. Amiri IS, Ali J (2014) Multiplex and De-multiplex of generated multi optical soliton by MRRs using fiber optics transmission link. Quantum Matter
93. Yupapin PP et al (2010) New communication bands generated by using a soliton pulse within a resonator system. Circuits Syst 1(2):71–75
94. Sadegh Amiri I et al (2013) Generation of potential wells used for quantum codes transmission via a TDMA network communication system. Secur Commun Netw 6(11):1301–1309
95. Amiri IS, Ali J (2012) Generation of nano optical tweezers using an Add/drop interferometer system. In: 2nd postgraduate student conference (PGSC), Singapore
96. Alavi SE et al (2014) Optical amplification of tweezers and bright soliton using an interferometer ring resonator system. J Comput Theor Nanosci (CTN)
97. Amiri IS et al (2012) A Study of dynamic optical tweezers generation for communication networks. Int J Adv Eng Technol (IJAET) 4(2):38–45
98. Ali J et al (2011) Dark soliton array for communication security. In: AMN-APLOC international conference, Wuhan, China
99. Amiri IS, Alavi SE, Ali J (2013) High capacity soliton transmission for indoor and outdoor communications using integrated ring resonators. Int J Commun Syst
100. Afroozeh A et al (2010) Optical dark and bright soliton generation and amplification. AIP Conf Proc 1341:259–263
101. Amiri IS, Ali J (2014) Deform of biological human tissue using inserted force applied by optical tweezers generated By PANDA ring resonator. Quantum Matter 3(1):24–28

102. Nikoukar A, Amiri IS, Ali J (2013) Generation of nanometer optical tweezers used for optical communication networks. Int J Innovative Res Comput Commun Eng 1(1):77–85
103. Amiri IS, Ali J (2013) Nano Optical tweezers generation used for heat surgery of a human tissue cancer cells using add/drop interferometer system. Quantum Matter 2(6):489–493
104. Amiri IS et al (2014) Soliton generation by ring resonator for optical communication application. In: Ramirez J (ed). Nova Science Publishers, Hauppauge, NY, 11788, USA
105. Ali J et al (2010) Novel system of fast and slow light generation using micro and nano ring resonators. In: International Conference ICAMN, Prince Hotel, Kuala Lumpur, Malaysia
106. Afroozeh A et al (2012) Simulation of soliton amplification in micro ring resonator for optical communication. Jurnal Teknologi (Sci Eng) 55:271–277

Chapter 2
Theoretical Background of Microring Resonator Systems and Soliton Communication

Optical network has been recognized as an efficiently network due to the good channel spacing and large bandwidths. However, the search for a new technique remains. Therefore in this book, an interesting technique using the dark soliton to generate new secured communication carriers is proposed [1, 2]. Dark soliton detection is difficult because its amplitude is vanished during the propagation. To date, several papers have investigated the dark soliton behaviors, where one of them has shown a fascinating result in the dark soliton can be converted into bright soliton and finally detected [3–5]. This means that dark soliton behavior can be used as an communication carrier so that it can be retrieved by the dark-bright soliton conversion. Here the dynamic trapping tool array (DTA) generation scheme is proposed, which are generated by using the multiplexed dark soliton pulses by using a pumped laser in a fiber optic system. The high capacity network and the security concept using multi dark soliton is discussed [6–8].

2.1 Evaluation of Soliton

The nonlinear Schrödinger equation (NLSE) is an appropriate equation for describing the propagation of light in optical fibers using normalization parameter such as: the normalized time T_0, the dispersion length L_D and peak power of the pulse P_0 the nonlinear Schrödinger equation in the terms of normalized coordinates can be written as:

$$i(\frac{\partial u}{\partial z}) - \frac{5}{2}(\frac{\partial^2 u}{\partial t^2}) + N^2 |u|^2 u + i(\frac{\alpha}{2})u = 0 \qquad (2.1)$$

© The Author(s) 2015
I. Sadegh Amiri et al., *Soliton Coding for Secured Optical Communication Link*,
SpringerBriefs in Applied Sciences and Technology,
DOI 10.1007/978-981-287-161-9_2

Fig. 2.1 Evolution of soliton in normal dispersion regime

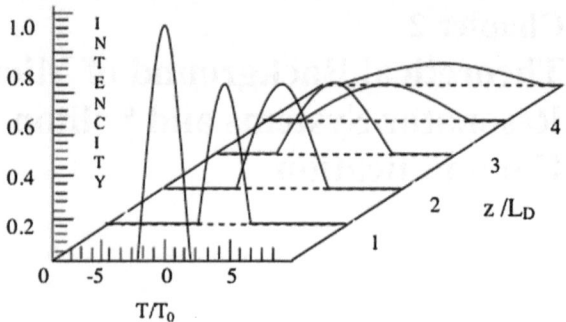

Fig. 2.2 Evolution of soliton in anomalous dispersion regime

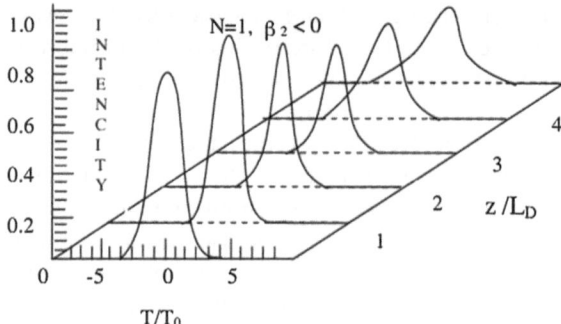

where u(z, t) is pulse envelope function, z is propagation distance along the fiber, N is an integer designating the order of soliton and α is the coefficient of energy gain per unit length, and with negative value it represents energy loss. Here s is −1 for negative β_2 (anomalous GVD-Bright soliton) and +1 for positive β_2 (normal GVD-Dark soliton) as shown in Figs. 2.1 and 2.2,

$$N_2 = \frac{L_D}{L_{NL}} = \frac{\gamma P_0 T_0^2}{|\beta_2|^2} \tag{2.2}$$

With nonlinear parameter γ and nonlinear length L_{NL}.

It is apparent that SPM dominates for N > 1 while for N < 1 dispersion effects dominates. For N ≈ 1 both SPM and GVD cooperate in such a way that the SPM-induced chirp is just right to cancel the GVD induced broadening of the pulse. The optical pulse would then propagate undistorted in the form of soliton. By integrating the NLS, the solution for the fundamental soliton can be written as

$$u(z, t) = \sec h(t) \exp(iz/2) \tag{2.3}$$

where, sec h(t) is hyperbolic scent function. Since the phase term exp(iz/2) has no influence on the shape of the pulse, the soliton is independent of z and hence is non dispersive in time domain. It is property of a fundamental soliton makes it an ideal candidate for optical communications. Optical solitons are very stable against perturbations; therefore they can be created even when the pulse shape

Fig. 2.3 Schematic diagram for a ring resonator coupled to a single waveguide

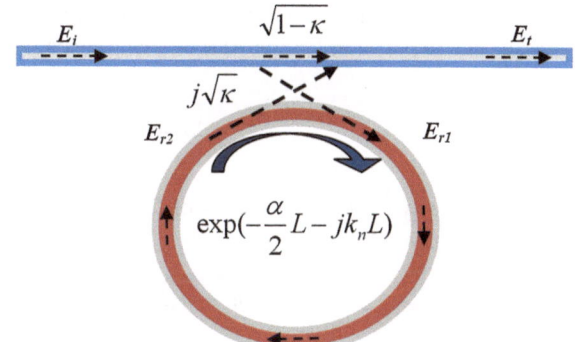

and peak power deviates from ideal conditions (values corresponding to N = 1). Soliton for Secured Communication.

To have the secured communication, the performance of resonators should be considered in terms of resonance width, the free spectral range, the finesse, and the quality factor. Expression for these characterizing quantities will be described as follows: Microring Resonator (MRR) Used to Generate Logic Codes:

A fiber optic ring resonator consists of a waveguide in a closed loop which is coupled to one or more input/output (or bus) waveguides [9–11]. A simple microring resonator is shown in Fig. 2.3.

2.2 Microring Resonator Used to Generate Logic Codes

Ring resonator provides traveling wave procedure, unlike the standing wave characteristic of Fabry-Perot resonators (F-P). Ring resonator can be considered as an interferometer device, which resonates for light whose phase change is an integer multiple of 2π after each trip around the ring [12–14]. Part of light that does not contribute this resonant condition will be transmitted through the bus waveguide. **Signal loss** occurs when light is transmitted through the fiber, especially over long distances such as undersea cables. The expression for the resonant wavelengths of the ring is very similar to that of the F-P and is given by [15, 16]

$$\lambda_r = \frac{2\pi R n_{eff}}{m} \tag{2.4}$$

where, R is the ring radius constructed with circular waveguide and m is an integer. In this situation the device will act as a phase filter where all wavelengths are transmitted and the resonant wavelengths, having also traversed the ring, acquires a phase change. To capture or separate the resonant wavelengths from the rest, an additional waveguide as an output bus, can be positioned on the opposite side of the ring. In this case the ring resonator is known as an add/drop filter system. The key performance parameters of the ring resonator include the *FSR*, the *ER*, and the finesse. The expression for the *FSR* of a ring resonator is given by

$$\Delta\lambda = \frac{\lambda_r^2}{2\pi R n_g} \tag{2.5}$$

The nonlinearity of the fiber ring is of the Kerr-type, wherein the nonlinear refractive index is given by

$$n = n_0 + n_2 I = n_0 + (\frac{n_2}{A_{eff}})P, \tag{2.6}$$

where n_0 and n_2 are the linear and nonlinear refractive indices [17, 18], while I and P are the optical intensity and optical field power, respectively. Here, the fiber coupler is considered as a point device and is reciprocal. The linear and nonlinear phase shifts of the ring resonator can be expressed by $\phi_0 = kLn_0$ and $\phi_{NL} = kLn_2|E_1|^2$, where $k = 2\pi/\lambda$ is a wave number, and $L = 2\pi R$ is the circumference of the ring resonator, where R is the radius of the ring resonator [19–22]. Mathematically, the subsequence equations of the round-trip within the system is given by [23, 24].

$$E_{n+1} = j\sqrt{(1-\gamma)\kappa} E_{in} + \sqrt{(1-\gamma)(1-\kappa)} x E_n \exp(-j(\phi_0 + \phi_{NL})) \tag{2.7}$$

Here, the subscript n denotes the number of round-trips inside the system. This equation has to be satisfied with boundary conditions appropriate for ring. The transmission around the single ring resonator is represented by

$$z^{-1} = \exp(-\alpha L/2 - jk_n L) \tag{2.8}$$

where k_n is the propagation constant and $\alpha L/2$ is the ring loss (round-trip loss), which includes propagation loss, losses resulting from transitions in the curvature, and bending losses. The value of α (unit length^{-1}) depends on the properties of the material and the waveguide used, and it is referred to as the intensity attenuation coefficient, where L is the circumference of the ring resonator. In order to define this, we consider a ring resonator connected to a single coupler that extracts light from the ring into the output waveguides.

When an input electric field, E_i is coupled to the ring waveguide through an external bus waveguide, a positive feedback is induced and the field inside the ring resonator, E_{r2} starts to build up. The feedback mechanism will be induced by the ring waveguide, therefore does not need any further requirements such as Bragg gratings, mirrors, or distributed feedback waveguides with difficult fabrication process. Due to on-resonant certain wavelength of the input signals inside the ring waveguide, frequency selectivity is obtained [25, 26]. The inserted and transmitted electric fields into the ring resonator are expressed by

$$E_{r1} = (1-\gamma)^{\frac{1}{2}}\left[jE_i\sqrt{\kappa} + E_{r2}\sqrt{1-\kappa}\right] \tag{2.9}$$

$$E_{r2} = E_{r1}\exp(-\frac{\alpha}{2}L - jk_n L) \tag{2.10}$$

where $k_n = \frac{2\pi \cdot n_{eff}}{\lambda}$ and γ denotes the intensity insertion loss coefficient of the directional coupler and n_{eff} is the effective refractive index. Therefore, the refractive index n quantifies the increase in the wave number (phase change per unit length) caused by the medium. Here, the effective refractive index n_{eff} has the similar meaning with light propagation in a waveguide, where it depends not only on the wavelength but also on the mode, in which the light propagates. The ratio of the output and input powers which is E_t/E_i can be calculated as [27].

$$\frac{E_t}{E_i} = (1-\gamma)^{\frac{1}{2}} \cdot \left[\frac{\sqrt{1-\kappa} - (1-\gamma)^{\frac{1}{2}} \cdot \exp(-\frac{\alpha}{2}L - jk_nL)}{1 - (1-\gamma)^{\frac{1}{2}} \cdot \sqrt{1-\kappa} \cdot \exp(-\frac{\alpha}{2}L - jk_nL)} \right] \qquad (2.11)$$

In the following new parameter will be used for simplifying [28]:

$$D = (1-\gamma)^{\frac{1}{2}}, \ x = D \cdot \exp(-\frac{\alpha}{2} \cdot L), \ y = \sqrt{1-\kappa}, \ \phi = k_nL$$

Intensity relation to the output port is given by [29]:

$$T = \frac{I_t}{I_i}(\varphi) = \left|\frac{E_t}{E_i}\right|^2 = D^2 \cdot \left[1 - \frac{(1-x^2) \cdot (1-y^2)}{(1-xy)^2 + 4xy \cdot \sin^2(\frac{\varphi}{2})} \right] \qquad (2.12)$$

Maximum and minimum transmission can be calculated when $\sin^2(\frac{\varphi}{2})$ is "1" and "0" respectively. Therefore;

$$T\text{max} = D^2 \cdot \frac{(x+y)^2}{(1+x \cdot y)^2} \qquad (2.13)$$

$$T\text{min} = D^2 \cdot \frac{(x-y)^2}{(1-x \cdot y)^2} \qquad (2.14)$$

The minimum transmission, $T_{\min}$ occurs at the resonant point when the circumference of the ring L, is an integer number of the guide wavelength, which is given by

$$\phi = k_n \cdot L = 2m\pi, \quad m = \text{integer,}$$
$$m \cdot \lambda_m = n \cdot L \qquad (2.15)$$

Here, m is the mode number, λ_m is the resonant mode wavelength. The on-off ratio for the single ring resonator is defined as the ratio of the on-resonance intensity to the off-resonance intensity which is maximum at $T_{\min} = 0$. Therefore $x = y$ and

$$\alpha = -\frac{1}{L} \times \ln(\frac{1-\kappa}{D^2}) \qquad (2.16)$$

This relationship given by Eq. (3.26) is also referred to as critical coupling, where the maximum on-off ratio $\frac{I_t}{I_i}(2m\pi) = 0$ can be obtained by varying the coupling coefficient (κ) or the intensity attenuation coefficient (α).

2.3 Resonance Bandwidth

Resonance bandwidth determines how fast optical data can be processed by a ring resonator. The resonator bandwidth is give by the full-width at half-maximum (FWHM or 3 dB bandwidth) $\delta\Phi[I_t/I_i(\varphi) = 0.5]$ and the finesse F of the resonator is given by:

$$\delta\phi = \frac{2(1 - xy)}{\sqrt{xy}} \tag{2.17}$$

To understand how the bandwidth of the resonator is affected by the coupling coefficient κ, we will consider critically coupled ring resonator. In such a case,

$$\delta\phi = \frac{2\kappa}{\sqrt{1 - \kappa}} \tag{2.18}$$

Therefore, the lower coupling coefficient, the smaller resonance bandwidth is obtained.

2.4 Finesse

The finesse of the resonator is defined as a ratio of the free spectral range (FSR) and the full width at half maximum of the resonance [30–32]. For the Fig. 2.4 using FSR (frequency spacing between two resonance) in terms of the is equal to 2π and thus the finesse is given by

$$F = \frac{2\pi}{\delta\phi} = \frac{\pi\sqrt{xy}}{(1 - xy)} \tag{2.19}$$

Fig. 2.4 Transmission characteristic of single ring resonator [33]

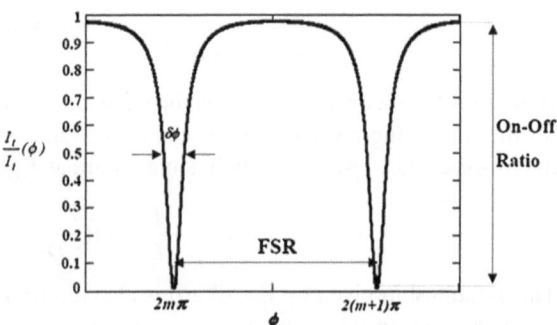

2.5 Free Spectral Range

The frequency spacing between two resonance peaks is called the free spectral range which can be calculated. The phase constant which corresponds to $\Phi = 2(m + 1)\pi$ is defined as κ. The phase constant corresponds to $\Phi = 2(m + 1)\pi$ is defined as $\kappa + \Delta\kappa$. The frequency shift Δf and the wavelength shift $\Delta\lambda$ are related to the variation of the phase constant $\Delta\kappa$ as $\Delta f = (c/2\pi) \cdot \Delta\kappa$ and $\Delta\lambda = -(\lambda^2/2\pi) \cdot \Delta\kappa$. The resonance spacing in terms of the frequency f and the wavelength λ are given by

$$\Delta f = \frac{c}{n_{gr} \cdot L} \tag{2.20}$$

$$\Delta\lambda = |-\frac{\lambda^2}{n_{gr} \cdot L}| \tag{2.21}$$

where n_{gr} is the group refractive index, which is defined as;

$$n_{gr} = n_{eff} - |\lambda\frac{dn_{eff}}{d\lambda}| \tag{2.22}$$

2.6 Quality Factor

Another value for characterization of ring resonator is the Q factor, The Q factor of the resonator is a measure of the sharpness of the resonance. In analogy with electrical circuit, the quality factor of an optical waveguide due it stored energy and the power lost per optical cycle. The Q factor is defined as

$$Q = \omega\frac{stored\ energy}{PowerLoss} \tag{2.23}$$

where ω is the frequency of the light coupled to the resonator. The Q factor of the resonator can be calculated from [34].

$$Q = \frac{f_0}{\delta f} = \frac{\lambda_0}{\delta\lambda} \tag{2.24}$$

The Q factor is the ratio of the absolute frequency f_0 or absolute wavelength λ_0 to the 3 dB bandwidth (δf or $\delta\lambda$). The shape and the bandwidth of the fiber response is determined by Q factor. The finesse and the Q factor are both important when one is interested in both the FSR (Δf or $\Delta\lambda$) and the 3 dB bandwidth (δf or $\delta\lambda$). They are related by:

$$\frac{Q}{F} = \frac{f_0}{\Delta f} = \frac{\lambda_0}{\Delta\lambda} \tag{2.25}$$

Fig. 2.5 Factor depending on the finesse for a specific radius R [35]

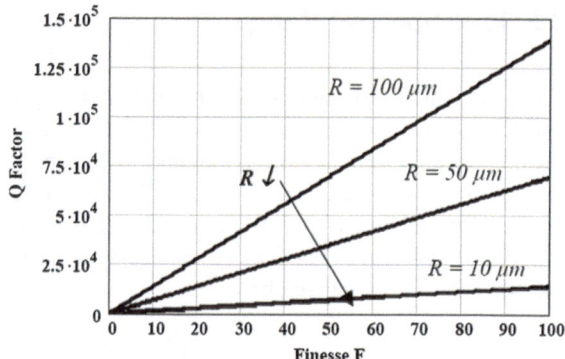

The Q factor depending on the finesse F for a ring resonator with a radius $R = 100$, 50 and 10 µm, a group refractive index $n_{gr} = 3.44$ at a wavelength of $\lambda = 1.55$ µm is shown in Fig. 2.5.

2.7 Add/Drop Filter System

Recently, optical ring resonators (ORR) have numerous applications in single mode lasers, biosensors, optical switching, add/drop filters, tunable lasers, signal processing and dispersion compensators [36–38]. In any WDM system, optical filters are used for separating one optical channel from the combined signals. The basic ORR with two couplers is illustrated in Fig. 2.6. The main performance characteristics of these resonators are the transmittance, free spectral range, finesse, Q-factor, and the group delay, which have been demonstrated both theoretically and experimentally in many works. Structural design of a single ring

Fig. 2.6 Ring resonator with two adjacent waveguide

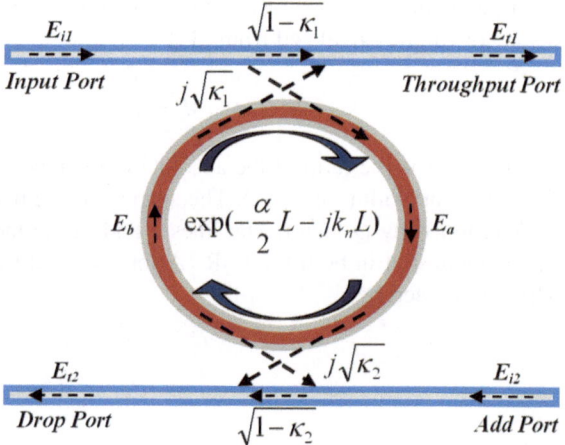

resonator (SRR) add/drop filter system is shown in Fig. 2.3, which is constructed by 2×2 optical couplers.

For simplification, the intensity relation [39] does not take into account coupling loses ($D^2 = 1$).

$$E_a = E_i 1 j \sqrt{\kappa_1} + E_b \sqrt{1 - \kappa_1} e^{\frac{-\alpha}{2} \frac{L}{2} - jk_n \frac{L}{2}} \tag{2.26}$$

$$E_b = E_a \sqrt{1 - \kappa_2} e^{\frac{-\alpha}{2} \frac{L}{2} - jk_n \frac{L}{2}} \tag{2.27}$$

$$E_a = \frac{E_i 1 j \sqrt{\kappa_1}}{1 - \sqrt{1 - \kappa_1}\sqrt{1 - \kappa_2} e^{\frac{-\alpha}{2} L - jk_n L}} \tag{2.28}$$

$$E_b = \frac{E_i 1 j \sqrt{\kappa_1}}{1 - \sqrt{1 - \kappa_1}\sqrt{1 - \kappa_2} e^{\frac{-\alpha}{2} L - jk_n L}} \cdot \sqrt{1 - \kappa_2} e^{\frac{-\alpha}{2} \frac{L}{2} - jk_n \frac{L}{2}} \tag{2.29}$$

$$E_{t1} = E_b e^{\frac{-\alpha}{2} \frac{L}{2} - jk_n \frac{L}{2}} j \sqrt{\kappa_1} + E_{i1} \sqrt{1 - \kappa_1} \tag{2.30}$$

$$E_{t2} = E_a e^{\frac{-\alpha}{2} \frac{L}{2} - jk_n \frac{L}{2}} j \sqrt{\kappa_2} \qquad at \quad E_{i2} = 0 \tag{2.31}$$

$$\begin{aligned}\frac{E_{t1}}{E_{i1}} &= \frac{-\kappa_1 \sqrt{1 - \kappa_2} e^{\frac{-\alpha}{2} L - jk_n L} + \sqrt{1 - \kappa_1} - (1 - \kappa_1)\sqrt{1 - \kappa_2} e^{\frac{-\alpha}{2} L - jk_n L}}{1 - \sqrt{1 - \kappa_1}\sqrt{1 - \kappa_2} e^{\frac{-\alpha}{2} L - jk_n L}} \\ &= \frac{-\sqrt{1 - \kappa_2} e^{\frac{-\alpha}{2} L - jk_n L} + \sqrt{1 - \kappa_1}}{1 - \sqrt{1 - \kappa_1}\sqrt{1 - \kappa_2} e^{\frac{-\alpha}{2} L - jk_n L}}\end{aligned} \tag{2.32}$$

$$\frac{E_{t2}}{E_{i1}} = \frac{-\sqrt{\kappa_1.\kappa_2} e^{\frac{-\alpha}{2} \frac{L}{2} - jk_n \frac{L}{2}}}{1 - \sqrt{1 - \kappa_1}\sqrt{1 - \kappa_2} e^{\frac{-\alpha}{2} L - jk_n L}} \tag{2.33}$$

where κ_1 and κ_2 are the coupling coefficients, $L = 2\pi R$ and R is the radius of the add/drop filter device [40, 41]. The normalized outputs of the add/drop filter system are expressed as [42]:

$$\frac{I_{t1}}{I_{i1}} = \left|\frac{E_{t1}}{E_{i1}}\right|^2 = \frac{1 - \kappa_1 - 2\sqrt{1 - \kappa_1}\sqrt{1 - \kappa_2} e^{\frac{-\alpha}{2} L} \cos(k_n L) + (1 - \kappa_2)e^{-\alpha L}}{1 + (1 - \kappa_1)(1 - \kappa_2)e^{-\alpha L} - 2\sqrt{1 - \kappa_1}\sqrt{1 - \kappa_2} e^{\frac{-\alpha}{2} L} \cos(k_n L)} \tag{2.34}$$

$$\frac{I_{t2}}{I_{i1}} = \left|\frac{E_{t2}}{E_{i1}}\right|^2 = \frac{\kappa_1 \cdot \kappa_2 e^{\frac{-\alpha}{2} L}}{1 + (1 - \kappa_1)(1 - \kappa_2)e^{-\alpha L} - 2\sqrt{1 - \kappa_1}\sqrt{1 - \kappa_2} e^{\frac{-\alpha}{2} L} \cos(k_n L)} \tag{2.35}$$

Using $y_1 = \sqrt{1 - \kappa_1}$ and $y_2 = \sqrt{1 - \kappa_2}$, the intensity relations are then given by:

$$\frac{I_{t1}}{I_{i1}}(\varphi) = \left|\frac{E_{t1}}{E_{i1}}\right|^2 = 1 - \frac{(1 - y_1^2) \cdot (1 - y_2^2 x^2)}{(1 - y_1 y_2 x)^2 + 4 y_1 y_2 x \sin^2\left(\frac{\varphi}{2}\right)} \tag{2.36}$$

$$\frac{I_{t2}}{I_{i1}}(\varphi) = \left|\frac{E_{t2}}{E_{i1}}\right|^2 = \frac{(1 - y_1^2) \cdot (1 - y_2^2) \cdot x}{(1 - y_1 y_2 x)^2 + 4 y_1 y_2 x \sin^2\left(\frac{\varphi}{2}\right)} \tag{2.37}$$

The full-width at half-maximum (FWHM) is given in this configuration by:

$$\delta\phi = 2\frac{1 - y_1 y_2 x}{\sqrt{y_1 y_2 x}}, \tag{2.38}$$

where the finesse F is given by:

$$F = \frac{2\pi}{\delta\phi} = \frac{\pi \sqrt{y_1 y_2 x}}{1 - y_1 y_2 x} \tag{2.39}$$

The maximum and minimum transmission are calculated as follows. For the throughput port:

$$T_{\max} = \frac{(y_1 + y_2 x)^2}{(1 + y_1 y_2 x)^2} \tag{2.40}$$

$$T_{\min} = \frac{(y_1 - y_2 x)^2}{(1 - y_1 y_2 x)^2} \tag{2.41}$$

And for the drop port:

$$T_{\max} = \frac{(1 - y_1^2) \cdot (1 - y_2^2) \cdot x}{(1 - y_1 y_2 x)^2} \tag{2.42}$$

$$T_{\min} = \frac{(1 - y_1^2) \cdot (1 - y_2^2) \cdot x}{(1 + y_1 y_2 x)^2} \tag{2.43}$$

The on-off ratio of an add/drop filter system is given by:

$$\frac{T_{\max}(through\,put\,port)}{T_{\min}(drop\,port)} = \frac{(y_1 + y_2 x)^2}{(1 - y_1^2) \cdot (1 - y_2^2) \cdot x} \tag{2.44}$$

The output intensity, I_{t1} at the throughput port will be zero at resonance ($k_n L = 2m\pi$) which indicates that the resonance wavelength is fully extracted by the resonator when $\kappa_1 = \kappa_2$ and $\alpha = 0$. The loss of signal power resulting from the insertion of a device in a transmission line for example an optical fiber is defined insertion loss and usually expressed in dBs. Therefore, it is a measure of attenuation. Attenuation can include loss due to the source and load impedances not matching, but is not included in insertion loss since this is a loss that was already present before the "insertion" was made. If the power transmitted to the load before insertion is P_T and the power received by the load after insertion is P_R, then the insertion loss in dB is given by,

$$IL = 10 \log_{10} \frac{P_T}{P_R} \tag{2.45}$$

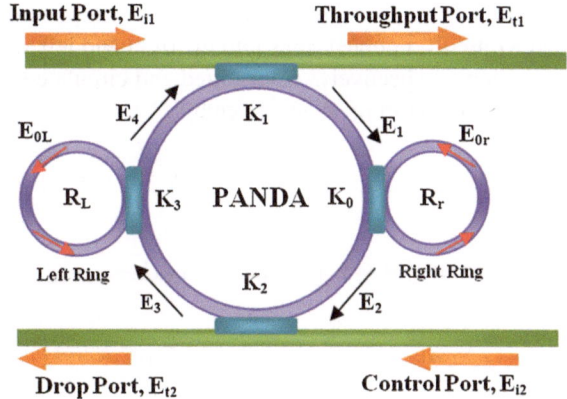

Fig. 2.7 Schematic of a PANDA ring resonator system

2.8 PANDA Ring Resonators

This system consists of one add/drop interferometer system connected to two ring resonators in the left and right sides. This system represents a new technique of combination and integration of micro ring resonators in which it can be widely used to improve the secure communication and the high capacity of optical signal proceeding in network communications [43]. Here the derived equations of the system is introduced which show that how does the input pulse propagates inside the rings systems [44–46]. The proposed system is shown in Fig. 2.7.

The resonator output fields, E_{t1} and E_1 consist of the transmitted and circulated components within the add/drop optical filter system, given by [47, 48]

$$E_{t1} = \sqrt{1 - \gamma_1}\left[\sqrt{1 - \kappa_1}E_{i1} + j\sqrt{\kappa_1}E_4\right], \tag{2.46}$$

$$E_1 = \sqrt{1 - \gamma_1}\left[\sqrt{1 - \kappa_1}E_4 + j\sqrt{\kappa_1}E_{i1}\right], \tag{2.47}$$

$$E_2 = E_{0r}E_1 e^{-\frac{\alpha}{2}\frac{L}{2} - jk_n\frac{L}{2}}, \tag{2.48}$$

where κ_1 is the intensity coupling coefficient, γ_1 is the fractional coupler intensity loss, α is the attenuation coefficient, $k_n = \frac{2\pi}{\lambda}$ is the wave propagation number, λ is the input wavelength light field, $L = 2\pi R_{ad}$ and R_{ad} is the radius of the add/drop system. For the second coupler of the add/drop system,

$$E_{t2} = \sqrt{1 - \gamma_2}\left[\sqrt{1 - \kappa_2}E_{i2} + j\sqrt{\kappa_2}E_2\right], \tag{2.49}$$

$$E_3 = \sqrt{1 - \gamma_2}\left[\sqrt{1 - \kappa_2}E_2 + j\sqrt{\kappa_2}E_{i2}\right], \tag{2.50}$$

$$E_4 = E_{0L}E_3 e^{-\frac{\alpha}{2}\frac{L}{2} - jk_n\frac{L}{2}}. \tag{2.51}$$

E_{0r} and E_{0L} are the light fields circulated components of the nanoring radii and R_r and R_L are the coupled rings into the right and left sides of the add/drop optical filter system, respectively. Transmitted and circulated components of the light fields in the right nanoring, R_r are given by

$$E_2 = \sqrt{1-\gamma}\left[\sqrt{1-\kappa_0}E_1 + j\sqrt{\kappa_0}E_{r2}\right], \tag{2.52}$$

$$E_{r1} = \sqrt{1-\gamma}\left[\sqrt{1-\kappa_0}E_{r2} + j\sqrt{\kappa_0}E_1\right], \tag{2.53}$$

$$E_{r2} = E_{r1}e^{-\frac{\alpha}{2}L_1 - jk_nL_1}. \tag{2.54}$$

or

$$E_{r1} = \frac{j\sqrt{1-\gamma}\sqrt{\kappa_0}\,E_1}{1-\sqrt{1-\gamma}\sqrt{1-\kappa_0}e^{-\frac{\alpha}{2}L_1 - jk_nL_1}}, \tag{2.55}$$

$$E_{r2} = \frac{j\sqrt{1-\gamma}\sqrt{\kappa_0}E_1e^{-\frac{\alpha}{2}L_1 - jk_nL_1}}{1-\sqrt{1-\gamma}\sqrt{1-\kappa_0}e^{-\frac{\alpha}{2}L_1 - jk_nL_1}}, \tag{2.56}$$

where $L_1 = 2\pi R_r$ and R_r is the radius of the right side nanoring. Thus, the output circulated light field [49–51], E_{0r}, for the right side nanoring is given by

$$E_{0r} = E_1\frac{\sqrt{(1-\gamma)(1-\kappa_0)} - (1-\gamma)e^{-\frac{\alpha}{2}L_1 - jk_nL_1}}{1-\sqrt{1-\gamma}\sqrt{1-\kappa_0}e^{-\frac{\alpha}{2}L_1 - jk_nL_1}}. \tag{2.57}$$

Similarly, the output circulated light field, E_{0L}, for the left side nanoring of the add/drop system is given by

$$E_{0L} = E_3\frac{\sqrt{(1-\gamma_3)(1-\kappa_3)} - (1-\gamma_3)e^{-\frac{\alpha}{2}L_2 - jk_nL_2}}{1-\sqrt{1-\gamma_3}\sqrt{1-\kappa_3}e^{-\frac{\alpha}{2}L_2 - jk_nL_2}}, \tag{2.58}$$

where $L_2 = 2\pi R_L$ and R_L is the radius of the left side nanoring. Regarding to more simplification such as $x_1 = (1-\gamma_1)^{1/2}$, $x_2 = (1-\gamma_2)^{1/2}$, $y_1 = (1-\kappa_1)^{1/2}$, and $y_2 = (1-\kappa_2)^{1/2}$, the interior circulated light fields, E_1, E_3 and E_4 are given by [52, 53].

$$E_1 = \frac{jx_1\sqrt{\kappa_1}E_{i1} + jx_1x_2y_1\sqrt{\kappa_2}E_{0L}E_{i2}e^{-\frac{\alpha}{2}\frac{L}{2} - jk_n\frac{L}{2}}}{1 - x_1x_2y_1y_2E_{0r}E_{0L}e^{-\frac{\alpha}{2}L - jk_nL}}, \tag{2.59}$$

$$E_3 = x_2y_2E_{0r}E_1e^{-\frac{\alpha}{2}\frac{L}{2} - jk_n\frac{L}{2}} + jx_2\sqrt{\kappa_2}E_{i2}, \tag{2.60}$$

$$E_4 = x_2y_2E_{0r}E_{0L}E_1e^{-\frac{\alpha}{2}L - jk_nL} + jx_2\sqrt{\kappa_2}E_{0L}E_{i2}e^{-\frac{\alpha}{2}\frac{L}{2} - jk_n\frac{L}{2}}. \tag{2.61}$$

Thus, the throughput port (E_{t1}) output is expressed by

$$E_{t1} = AE_{i1} - BE_{i2}e^{-\frac{\alpha}{2}\frac{L}{2}-jk_n\frac{L}{2}} \left[\frac{CE_{i1}\left(e^{-\frac{\alpha}{2}\frac{L}{2}-jk_n\frac{L}{2}}\right)^2 + DE_{i2}\left(e^{-\frac{\alpha}{2}\frac{L}{2}-jk_n\frac{L}{2}}\right)^3}{1 - F\left(e^{-\frac{\alpha}{2}\frac{L}{2}-jk_n\frac{L}{2}}\right)^2} \right]$$

(2.62)

where,

$A = x_1x_2$, $B = x_1x_2y_2\sqrt{\kappa_1}E_{0L}$, $C = x_1^2x_2\kappa_1\sqrt{\kappa_2}E_{0r}E_{0L}$, $D = (x_1x_2)^2y_1y_2$ $\sqrt{\kappa_1\kappa_2}E_{0r}E_{0L}^2$ and $F = x_1x_2y_1y_2E_{0r}E_{0L}$. The power output of the throughput port (P_{t1}) is given by

$$P_{t1} = (E_{t1}) \cdot (E_{t1})^* = |E_{t1}|^2.$$

(2.63)

Similarly, the output optical field of the drop port (E_{t2}) is given by [54, 55].

$$E_{t2} = x_2y_2E_{i2} \left[\frac{x_1x_2\sqrt{\kappa_1\kappa_2}E_{0r}E_{i1}e^{-\frac{\alpha}{2}\frac{L}{2}-jk_n\frac{L}{2}} + x_1x_2^2y_1y_2\sqrt{\kappa_2}E_{0r}E_{0L}E_{i2}\left(e^{-\frac{\alpha}{2}\frac{L}{2}-jk_n\frac{L}{2}}\right)^2}{1 - x_1x_2y_1y_2E_{0r}E_{0L}\left(e^{-\frac{\alpha}{2}\frac{L}{2}-jk_n\frac{L}{2}}\right)^2} \right],$$

(2.64)

where the power output of the drop port (P_{t2}) is expressed by

$$P_{t2} = (E_{t2}) \cdot (E_{t2})^* = |E_{t2}|^2$$

(2.65)

Single ring resonator consists of a single coupler and a MRR. Schematic of the single ring resonator is illustrated in Fig. 2.6.

2.9 Multiple Narrow Pulse Switching Generation for Continuous Variable Quantum Key Distribution

Schematic diagram of the proposed system is as shown in Fig. 2.8. A soliton pulse with 20 ns pulse width, peak power at 500 mW is inserted into the system. The parameters of the system are fixed to $\lambda_0 = 1.55$ µm, $n_0 = 3.34$ (InGaAsP/InP), $A_{eff} = 0.50$, 0.25 and 0.12 µm^2 for the different radii of MRRs respectively, $\alpha = 0.5$ dB mm^{-1}, $\gamma = 0.1$. The coupling coefficient (kappa, κ) of the MRR ranged from 0.50 to 0.975 [56, 57].

Input optical field (E_{in}) of the bright soliton pulse can be expressed as,

$$E_{in} = A \sec h\left[\frac{T}{T_0}\right] \exp\left[\left(\frac{z}{2L_D}\right) - i\omega_0 t\right]$$

(2.66)

A and z are the optical field amplitude and propagation length, respectively. T is a soliton pulse propagation time in a form proceeding at the group speed [58], $T = t - \beta_1 \times z$, where β_1 and β_2 are the coefficients of the linear and second order terms of Taylor expansion of the propagation constant. $L_D = T_0^2/|\beta_2|$ is the

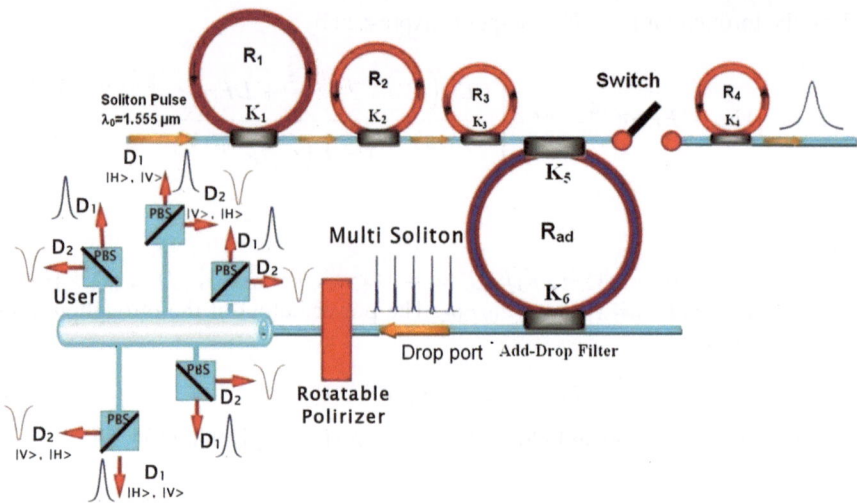

Fig. 2.8 Schematic diagram of a single and multiple narrow pulse switching generation for continuous variable quantum key distribution with the different time slot, where *PBS* polarizing beam splitter, *Ds* detectors, *Rs* ring radii and κs coupling coefficients

dispersion length of the soliton pulse. The frequency carrier of the soliton is w_0. When soliton peak intensity $\left(|\beta_2/\Gamma T_0^2|\right)$ is given, and then T_0 is known. The nonlinear length is given by $L_{NL} = 1/\Gamma\varphi_{NL}$ where $\Gamma = n_2 \times k_0$ is the length scale, thus $L_{NL} = L_D$ should be satisfied [59–61]. The refractive index (n) of light within the medium is given by Eq. (2.6) [62, 63].

The resonant output is formed, thus, the normalized output of the light field is the ratio between the output and input fields $E_{out}(t)$ and $E_{in}(t)$ in each roundtrip, which can be expressed as [64, 65]

$$|E_{out}(t)|^2 = |E_{in}(t)|^2 \times (1-\gamma) - \frac{(1 - x^2 + 2\gamma x^2 - \gamma - \gamma^2 x^2)\kappa}{(1 - x\sqrt{1-\gamma}\sqrt{1-\kappa})^2 + 4x\sqrt{1-\gamma}\sqrt{1-\kappa}\sin^2(\frac{\phi}{2})}.$$

$$(2.67)$$

Here the particular case of a Fabry-Perot cavity, which has an input and output mirror with a field reflectivity, $(1 - \kappa)$ [66–68], and a fully reflecting mirror is presented. κ is the coupling coefficient, and $x = \exp(-\alpha L/2)$ represents a roundtrip loss coefficient, $\Phi_0 = kLn_0$ and $\Phi_{NL} = kLn_2|E_{in}|^2$ are the linear and nonlinear phase shifts, $k = 2\pi/\lambda$ is the wave propagation [69, 70]. L and α are a waveguide length and linear absorption coefficient, respectively. The obtained results are based on solving the nonlinear Schrödinger Equation (NLSE) for the case of ring resonators using Optisystem and MATLAB programming.

2.10 Continuous Variable Quantum Key Distribution with the Different Time Slot Entangled Photon Encoding

The schematic diagram of the proposed system is as shown in Fig. 2.9. This system is associated with the practical device.

The soliton pulse is introduced into the proposed system. The input optical field (E_{in}) of the bright soliton pulse can be expressed by Eq. (2.66).

For a soliton pulse, there is a balance between dispersion and nonlinear lengths, hence $L_D = L_{NL}$ [71, 72]. When light propagates within the nonlinear medium, the refractive index (n) of light within the medium is given by Eq. (2.6) [73, 74].

The resonant output is can be formed, thus the normalized output of the light field is given by [75–77].

$$\left|\frac{E_{out}(t)}{E_{in}(t)}\right|^2 = (1-\gamma)\left[1 - \frac{(1-(1-\gamma)x^2)\kappa}{(1-x\sqrt{1-\gamma}\sqrt{1-\kappa})^2 + 4x\sqrt{1-\gamma}\sqrt{1-\kappa}\sin^2(\frac{\phi}{2})}\right]$$

(2.68)

where, the output and input fields in each round-trip are presented by $E_{out}(t)$ and $E_{in}(t)$. Equation (3) indicates that a ring resonator in the particular case is very similar to a Fabry-Perot cavity, which has an input and an output mirror with a field reflectivity, $(1-\kappa)$, and a fully reflecting mirror. κ is the coupling coefficient, and $x = \exp(-\alpha L/2)$ represents a round-trip loss coefficient, $\Phi_0 = kLn_0$ and $\Phi_{NL} = kLn_2|E_{in}|^2$ are the linear and nonlinear phase shifts, $k = 2\pi/\lambda$ is the wave propagation number in a vacuum. L and α are a waveguide length and linear absorption coefficient, respectively [78–80]. In this work, the iterative method is introduced to obtain the results.

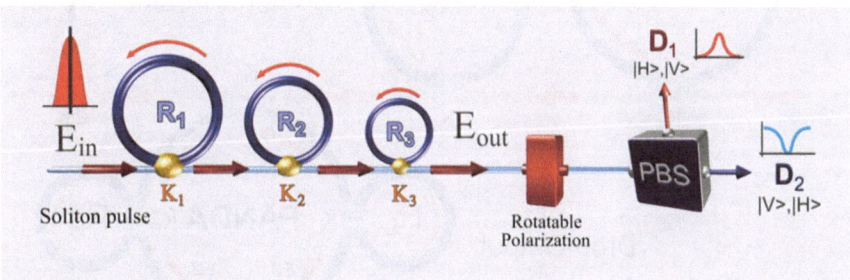

Fig. 2.9 Schematic diagram of a continuous variable quantum key distribution with the different time slot entangled photon encoding. *PBS* polarizing beam splitter, *Ds* detectors, *Rs* ring radii and *κs* coupling coefficients

2.11 GHz Soliton Carrier Generation for Transferring the Logic Codes Generated by the Chaotic Signals

The system of GHz frequency band generation is shown in Fig. 2.10. Here, series of ring resonators are connected to a panda ring resonator. The filtering process of the input soliton pulses is performed via the ring resonators, where frequency band ranges of 40–60 GHz can be obtained via the output signals of the panda ring resonator system.

Ring resonators are made of fibre optics where the medium has Kerr effect-type nonlinearity. The Kerr effect causes the refractive index (n) of the medium to vary as given in Eq. (2.6) [81].

A bright soliton with a central frequency of 50 GHz and power of 1 W is introduced into the first ring resonator, R_1, expressed by Eq. (2.66). The output and input signals in each round trip of the ring resonator can be expressed by Eq. (2.68). For the panda ring resonator, the interior signals are given as follows [82];

$$E_1 = \sqrt{1 - \gamma_3}\left(\sqrt{1 - \kappa_3}E_4 + j\sqrt{\kappa_3}E_{out2}\right) \tag{2.69}$$

and

$$E_2 = E_r E_1 e^{-\frac{\alpha}{2}\frac{L}{2} - jk_n\frac{L}{2}} \tag{2.70}$$

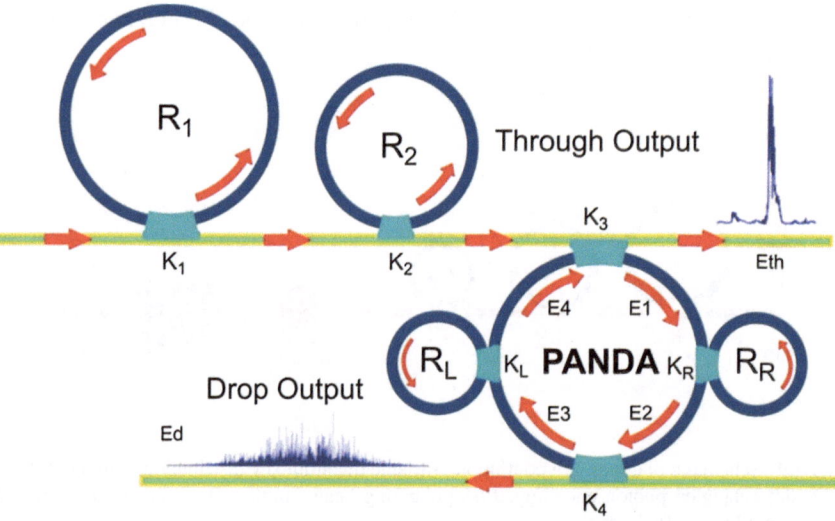

Fig. 2.10 Optical frequency band generation system using a panda ring resonator connected to a series of ring resonators

where κ_3 is the intensity coupling coefficient, γ_3 is the fractional coupler intensity loss, α is the attenuation coefficient, $L = 2\pi R_{PANDA}$, and R_{PANDA} is the radius of the panda system [83, 84].

The electric field of the right ring of the panda system is given by

$$E_r = E_1 \frac{\sqrt{(1 - \gamma_r)(1 - \kappa_r)} - (1 - \gamma_r)e^{-\frac{\alpha}{2}L_r - jk_nL_r}}{1 - \sqrt{1 - \gamma_r}\sqrt{1 - \kappa_r}e^{-\frac{\alpha}{2}L_r - jk_nL_r}}. \tag{2.71}$$

Inserting Eq. (6) into Eq. (5) results in

$$E_2 = \sqrt{1 - \gamma_r}\left(\sqrt{1 - \kappa_r}E_1 + j\sqrt{\kappa_r}E_{r2}\right), \tag{2.72}$$

where

$$E_{r2} = E_{r1}e^{-\frac{\alpha}{2}L_r - jk_nL_r} \tag{2.73}$$

and

$$E_{r1} = \sqrt{1 - \gamma_r}\left(\sqrt{1 - \kappa_r}E_{r2} + j\sqrt{\kappa_r}E_1\right). \tag{2.74}$$

E_{r1} and E_{r2} are round-trip light fields of the right ring and are given by

$$E_{r1} = \frac{j\sqrt{1 - \gamma_r}\sqrt{\kappa_r}E_1}{1 - \sqrt{1 - \gamma_r}\sqrt{1 - \kappa_r}e^{-\frac{\alpha}{2}L_r - jk_nL_r}} \tag{2.75}$$

and

$$E_{r2} = \frac{j\sqrt{1 - \gamma_r}\sqrt{\kappa_r}E_1 e^{-\frac{\alpha}{2}L_r - jk_nL_r}}{1 - \sqrt{1 - \gamma_r}\sqrt{1 - \kappa_r}e^{-\frac{\alpha}{2}L_r - jk_nL_r}}, \tag{2.76}$$

where κ_r is the intensity coupling coefficient, γ_r is the fractional coupler intensity loss, α is the attenuation coefficient, $k_n = \frac{2\pi}{\lambda}$, $L_r = 2\pi R_r$, and $R_r = 18$ μm is the radius of the right ring [85].

Input and output light fields of the left ring resonator can be expressed as

$$E_3 = \sqrt{1 - \gamma_4} \times \sqrt{1 - \kappa_4}E_2, \tag{2.77}$$

and

$$E_4 = E_l E_3 e^{-\frac{\alpha}{2}\frac{L}{2} - jk_n\frac{L}{2}} \tag{2.78}$$

where

$$E_l = E_3 \frac{\sqrt{(1 - \gamma_l)(1 - \kappa_l)} - (1 - \gamma_l)e^{-\frac{\alpha}{2}L_l - jk_nL_l}}{1 - \sqrt{1 - \gamma_l}\sqrt{1 - \kappa_l}e^{-\frac{\alpha}{2}L_l - jk_nL_l}}. \tag{2.79}$$

Here, $L_l = 2\pi R_L$, and $R_L = 8$ μm is the radius of left ring. Therefore, the output signals from the through and drop ports of the panda ring resonator can be expressed as

$$E_{th} = \sqrt{1-\gamma_3}\left[\sqrt{1-\kappa_3}E_{out2} + j\sqrt{\kappa_3}E_4\right] \tag{2.80}$$

and

$$E_d = \sqrt{1-\gamma_4} \times j\sqrt{\kappa_4}E_2. \tag{2.81}$$

In order to simplify these equations, the parameters of x_1, x_2, y_1 and y_2 are defined as

$$x_1 = (1-\gamma_3)^{\frac{1}{2}}, \ x_2 = (1-\gamma_4)^{\frac{1}{2}}, \ y_1 = (1-\kappa_3)^{\frac{1}{2}} \ \text{and} \ y_2 = (1-\kappa_4)^{\frac{1}{2}}.$$

Therefore,

$$E_1 = \frac{jx_1\sqrt{\kappa_3}E_{out2}}{1 - x_1x_2y_1y_2E_rE_le^{-\frac{\alpha}{2}L-jk_nL}} \tag{2.82}$$

$$E_2 = \frac{E_r \times jx_1\sqrt{\kappa_3}E_{out2} \times e^{-\frac{\alpha}{2}\frac{L}{2}-jk_n\frac{L}{2}}}{1 - x_1x_2y_1y_2E_rE_le^{-\frac{\alpha}{2}L-jk_nL}} \tag{2.83}$$

$$E_3 = x_2y_2E_rE_le^{-\frac{\alpha}{2}\frac{L}{2}-jk_n\frac{L}{2}} \tag{2.84}$$

$$E_4 = x_2y_2E_rE_lE_1e^{-\frac{\alpha}{2}L-jk_nL} \tag{2.85}$$

$$E_{th} = x_1(y_1E_{out2} + j\sqrt{\kappa_3}E_4) \tag{2.86}$$

$$E_d = x_2 \times j\sqrt{\kappa_4}E_2 \tag{2.87}$$

2.12 Digital Binary Codes Transmission via TDMA Networks Communication

Input optical field of dark soliton and Gaussian pulse are introduced into the input and add ports of the proposed add/drop interferometer system respectively, shown in Fig. 2.11.

Input optical fields of the dark soliton (E_{in}) and the Gaussian pulse (E_{add}) are expressed as Eqs. 1 and 2.

$$E_{in} = A \tan h\left[\frac{T}{T_0}\right] \exp\left[\left(\frac{z}{2L_D}\right) - i\omega_0 t\right] \tag{2.88}$$

Fig. 2.11 A schematic diagram of an add/drop interferometer system

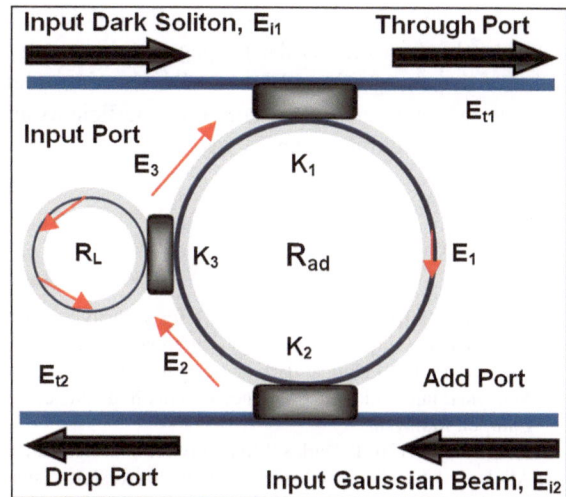

$$E_{add}(t) = E_0 \exp\left[\left(\frac{z}{2L_D}\right) - i\omega_0 t\right] \tag{2.89}$$

A and z are the optical field amplitude and propagation distance, respectively. T is a soliton pulse propagation time in a frame moving at the group velocity, $T = t - \beta_1 \times z$, where β_1 and β_2 are the coefficients of the linear and second order terms of Taylor expansion of the propagation constant. $L_D = T_0^2/|\beta_2|$ is the dispersion length of the soliton pulse. The frequency carrier of the soliton is ω_0. Solution is realized like pulse that keeps its temporal width invariance as it propagates, and thus is known as temporal soliton. Soliton peak intensity is $\left(|\beta_2/\Gamma T_0^2|\right)$. Here $\Gamma = n_2 \times k_0$, is the length scale over which dispersive or nonlinear effects makes the beam become wider or narrower. A balance should be achieved between the dispersion length (L_D) and the nonlinear length ($L_{NL} = (1/\gamma\varphi_{NL})$, where γ and φ_{NL} are the coupling loss of the field amplitude and nonlinear phase shift, thus $L_D = L_{NL}$ [86].

The iterative method is introduced to obtain the results. Add/drop interferometer system is proposed with appropriate parameters to generate optical tweezers in the form of potential wells. Two complementary optical circuits of the system can be given by [12, 87, 88]

$$\left|\frac{E_{t1}}{E_{in}}\right|^2 = \frac{(1-\kappa_1) - 2\sqrt{1-\kappa_1} \cdot \sqrt{1-\kappa_2}e^{-\frac{\alpha}{2}L}\cos(k_nL) + (1-\kappa_2)e^{-\alpha L}}{1 + (1-\kappa_1)(1-\kappa_2)e^{-\alpha L} - 2\sqrt{1-\kappa_1} \cdot \sqrt{1-\kappa_2}e^{-\frac{\alpha}{2}L}\cos(k_nL)}, \tag{2.90}$$

$$\left|\frac{E_{t2}}{E_{in}}\right|^2 = \frac{\kappa_1\kappa_2 e^{-\frac{\alpha}{2}L}}{1 + (1-\kappa_1)(1-\kappa_2)e^{-\alpha L} - 2\sqrt{1-\kappa_1} \cdot \sqrt{1-\kappa_2}e^{-\frac{\alpha}{2}L}\cos(k_nL)}, \tag{2.91}$$

where E_{t1} and E_{t2} represents the optical fields of the through and drop ports respectively. $\beta = kn_{eff}$ is the propagation constant, n_{eff} is the effective refractive index of the waveguide and $L = 2\pi R$ is the circumference of the MRR ring, where R is the radius. κ_1 and κ_2 are coupling coefficients and the ring resonator loss is α. The fractional coupler intensity loss is γ.

References

1. Ali J et al (2010) Generation of DSA for security application. In: 2nd international science, social science, engineering energy conference (I-SEEC 2010), Nakhonphanom, Thailand
2. Ali J et al (2010) Optical dark and bright soliton generation and amplification. In: Nanotech Malaysia, international conference on enabling science and technology 2010, KLCC, Kuala Lumpur, Malaysia
3. Amiri IS et al (2011) Dark soliton array for communication security. Proc Eng 8:417–422
4. Ali J et al (2010) Dark-bright solitons conversion system via an add/drop filter for signal security application. In: ICEM2010, Legend Hotel, Kuala Lumpur, Malaysia
5. Amiri IS et al (2012) Decimal convertor application for optical wireless communication by generating of dark and bright signals of soliton. Int J Eng Res Technol (IJERT) 1(5)
6. Afroozeh A et al (2011) Multi soliton generation for enhance optical communication. Appl Mech Mater 83:136–140
7. Ali J et al (2010) Multi-soliton generation and storage for nano optical network using nano ring resonators. In: ICAMN, international conference 2010, Prince Hotel, Kuala Lumpur, Malaysia
8. Ali J et al (2010) Multi-wavelength narrow pulse generation using MRR. In: ICAMN, international conference 2010, Prince Hotel, Kuala Lumpur, Malaysia
9. Afroozeh A et al (2012) Fast light generation using GaAlAs/GaAs waveguide. Jurnal Teknologi (Sci Eng) 57:17–23
10. Amiri IS, Nikoukar A, Ali J (2013) New system of chaotic signal generation based on coupling coefficients applied to an add/drop system. Int J Adv Eng Technol (IJAET) 6(1):78–87
11. Nikoukar A, Amiri IS, Ali J (2011) Secured binary codes generation for computer network communication. In: Network technologies and communications (NTC) conference 2010–2011, Singapore
12. Shahidinejad A, Sadegh Amiri I, Anwar T (2014) Enhancement of indoor WDM-based optical wireless communication using microring resonator. Rev Theor Sci 2(3):201–210
13. Amiri IS, Ali J (2013) Nano particle trapping by ultra-short tweezer and wells using MRR interferometer system for spectroscopy application. Nanosci Nanotechnol Lett 5(8):850–856
14. Amiri IS et al (2013) Optical stretcher of biological cells using sub-nanometer optical tweezers generated by an add/drop MRR system. Nanosci Nanotechnol Lett 6(2):111–117
15. Amiri IS, Ali J (2014) Femtosecond optical quantum memory generation using optical bright soliton. J Comput Theor Nanosci (CTN) 11(6):1480–1485
16. Amiri IS et al (2012) Generation of quantum codes using up and down link optical solition. Jurnal Teknologi (Sci Eng) 55:97–106
17. Amiri IS et al (2012) Generation of quantum photon information using extremely narrow optical tweezers for computer network communication. GSTF J Comput (joc) 2(1):140
18. Zeinalinezhad A et al (2014) Stop light generation using nano ring resonators for ROM. J Comput Theor Nanosci (CTN)
19. Afroozeh A et al (2010) Effect of center wavelength on MRR performance. In: Faculty of science postgraduate conference (FSPGC). Universiti Teknologi Malaysia
20. Amiri IS, Shahidinejad A, Ali J (2014) Generating of 57–61 GHz frequency band using a panda ring resonator. Quantum Matter

21. Amiri IS et al (2012) Molecular transporter system for qubits generation. Jurnal Teknologi (Sci Eng) 55:155–165
22. Amiri IS, Ali J (2014) Simulation of the single ring resonator based on the Z-transform method theory. Quantum Matter 3(6):519–522
23. Mohamad FK et al (2010) Finesse improvements of light pulses within MRR system. In: International conference on experimental mechanics (ICEM) 2010, Kuala Lumpur, Malaysia
24. Ali J et al (2010) Optical bistability in a FORR. In: ICEM 2010, Legend Hotel, Kuala Lumpur, Malaysia
25. Glaser W (1997) Photonik für Ingenieure. Verl. Technik
26. Yupapin P, Pornsuwancharoen N (2008) Guided wave optics and photonics: micro-ring resonator design for telephone network security. Nova Science Publ., New York
27. Amiri IS, Ali J (2014) Optical quantum generation and transmission of 57–61 GHz frequency band using an optical fiber optics. J Comput Theor Nanosci (CTN) 11(10)
28. Alavi SE et al (2014) Indoor data transmission over ubiquitous infrastructure of powerline cables and led lighting. J Comput Theor Nanosci (CTN)
29. Okamoto K (2006) Fundamentals of optical waveguides. Academic press, Massachusetts
30. Amiri IS, Ali J, Yupapin PP (2012) Enhancement of FSR and finesse using add/drop filter and PANDA Ring resonator systems. Int J Mod Phys B 26(04):1250034
31. Jalil MA et al (2010) Finesse improvements of light pulses within MRR system. In: Faculty of science postgraduate conference (FSPGC). Universiti Teknologi Malaysia, Malaysia
32. Ali J et al (2010) Narrow UV pulse generation using MRR and NRR system. In: ICAMN international conference 2010, Prince Hotel, Kuala Lumpur, Malaysia
33. Pornsuwanchareon PPYAN (2009) Guided wave optics and photonics. Nova science publisher, NewYork, 6 Dec 2009
34. Pornsuwanchareon PPYAN (2008) Guided eave optics and photonics: micro ring resonator design for telephone network security. Nova Science Publisher Inc. NewYork
35. Amiri IS et al (2013) All optical OFDM generation for IEEE802.11a based on soliton carriers using microring resonators. IEEE Photonics J 6(1)
36. Amiri IS et al (2014) Nanometer bandwidth soliton generation and experimental transmission within nonlinear fiber optics using an add-drop filter system. J Comput Theor Nanosci (CTN)
37. Amiri IS, Naraei P (2014) Optical transmission characteristics of an optical add-drop interferometer system. Quantum Matter
38. Yariv A (2000) Universal relations for coupling of optical power between microresonators and dielectric waveguides. Electron Lett 36(4):321–322
39. Amiri IS, Gifany D, Ali J (2013) Entangled photon encoding using trapping of picoseconds soliton pulse. IOSR J Appl Phys (IOSR-JAP) 3(1):25–31
40. Amiri IS et al (2014) The proposal of high capacity GHz soliton carrier signals applied for wireless commutation. Rev Theor Sci 2(4)
41. Nikoukar A et al (2012) MRR quantum dense coding for optical wireless communication system using decimal convertor. In: Computer and communication engineering (ICCCE) conference. IEEE Explore, Malaysia
42. Amiri IS, Afroozeh A, Bahadoran M (2011) Simulation and analysis of multisoliton generation using a PANDA ring resonator system. Chin Phys Lett 28(10):104205
43. Reece P, Wright E, Dholakia K (2007) Experimental observation of modulation instability and optical spatial soliton arrays in soft condensed matter. Phys Rev Lett 98(20):203902
44. Brambilla G et al (2007) Optical manipulation of microspheres along a subwavelength optical wire. Opt Lett 32(20):3041–3043
45. Shvedov VG et al (2010) Selective trapping of multiple particles by volume speckle field. Opt Express 18(3):3137–3142
46. Amiri IS et al (2013) IEEE 802.15.3c WPAN standard using millimeter optical soliton pulse generated by a panda ring resonator. IEEE Photonics J 5(5):7901912

47. Afroozeh A et al (2012) THz frequency generation using MRRs for THz imaging. In: International conference on enabling science and nanotechnology (EsciNano). IEEE Explore, Kuala Lumpur, Malaysia
48. Amiri IS et al (2012) Frequency-wavelength trapping by integrated ring resonators for secured network and communication systems. Int J Eng Res Technol (IJERT) 1(5)
49. Amiri IS (2014) Light detection and ranging using NIR (810 nm) laser source. In: Jian A (ed). LAP LAMBERT Academic Publishing, Germany
50. Amiri IS et al (2014) Optical stretcher of biological cells using sub-nanometer optical tweezers generated by an add/drop microring resonator system. Nanosci Nanotechnol Lett 6(2):111–117
51. Ali J et al (2010) Generation of continuous optical spectrum by soliton into a nano-waveguide. In: ICAMN international conference 2010, Prince Hotel, Kuala Lumpur, Malaysia
52. Ali J et al (2010) Localization of soliton pulse using nano-waveguide. In: ICAMN, international conference 2010, Prince Hotel, Kuala Lumpur, Malaysia
53. Amiri IS et al (2011) Generation of DSA for security application. Proc Eng 8:360–365
54. Kouhnavard M et al (2010) New system of chaotic signal generation using MRR. In: International conference on experimental mechanics (ICEM) 2010, Kuala Lumpur, Malaysia
55. Ali J et al (2010) Short and millimeter optical soliton generation using dark and bright soliton. In: AMN-APLOC international conference 2010, Wuhan, China
56. Ali J et al (2010) Effects of MRR parameter on the bifurcation behavior. In: Nanotech Malaysia, international conference on enabling science and technology 2010, KLCC, Kuala Lumpur, Malaysia
57. Amiri IS et al (2010) Storage of optical soliton wavelengths using NMRR. In: International conference on experimental mechanics (ICEM) 2010, Kuala Lumpur, Malaysia
58. Amiri IS et al (2012) Multi optical soliton generated by PANDA ring resonator for secure network communication. In: Computer and communication engineering (ICCCE) conference 2012. IEEE Explore
59. Amiri IS, Gifany D, Ali J (2013) Long distance communication using localized optical soliton via entangled photon. IOSR J Appl Phys (IOSR-JAP) 3(1):32–39
60. Alavi SE et al (2013) Optical wired/wireless communication using soliton optical tweezers. Life Sci J 10(12s):179–187
61. Ali J et al (2010) Stopping a dark soliton pulse within an NNRR. In: AMN-APLOC international conference 2010, Wuhan, China
62. Mohamad FK et al (2010) Effect of center wavelength on MRR performance. In: International conference on experimental mechanics (ICEM)2010, Kuala Lumpur, Malaysia
63. Ali J et al (2010) Wide and narrow signal generation using chaotic wave. In: Nanotech Malaysia, international conference on enabling science and technology 2010, Kuala Lumpur, Malaysia
64. Amiri IS, Ali J (2014) Picosecond soliton pulse generation using a PANDA system for solar cells fabrication. J Comput Theor Nanosci (CTN) 11(3):693–701
65. Ridha NJ et al (2010) Soliton signals and the effect of coupling coefficient in MRR systems. In: International conference on experimental mechanics (ICEM) 2010, Kuala Lumpur, Malaysia
66. Afroozeh A, Amiri IS, Ali J (2014) Slow light generation using micro waveguide. In: DSilva DL (ed). Springer, USA
67. Ali J et al (2010) Fast and slow lights via an add/drop device. In: ICEM 2010, Legend Hotel, Kuala Lumpur, Malaysia
68. Amiri IS et al (2014) Analytical treatment of the ring resonator passive systems and bandwidth characterization using directional coupling coefficients. J Comput. Theor. Nanosci. (CTN)
69. Ali J et al (2010) Security confirmation using temporal dark and bright soliton via nonlinear system. In: ICAMN, international conference 2010, Prince Hotel, Kuala Lumpur, Malaysia

70. Ali J et al (2010) Soliton wavelength division in MRR and NRR systems. In: AMN-APLOC international conference 2010, Wuhan, China
71. Ali J et al (2010) Narrow UV pulse generation using MRR and NRR system. In: ICAMN, international conference 2010, Prince Hotel, Kuala Lumpur
72. Amiri IS et al (2010) Storage of atom/molecules/photon using optical potential wells. In: The international conference on experimental mechanics (ICEM) 2010, Kuala Lumpur, Malaysia
73. Amiri IS, Nikoukar A, Ali J (2011) Quantum information generation using optical potential well. In: Network technologies and communications (NTC) conference 2010–2011, Singapore
74. Ali J et al (2010) Temporal dark soliton behavior within multi-ring resonators. In: Nanotech Malaysia, international conference on enabling science and technology 2010, Malaysia
75. Ridha NJ et al (2010) Controlling center wavelength and free spectrum range by MRR radii. In: International conference on experimental mechanics (ICEM) 2010, Kuala Lumpur, Malaysia
76. Ali J et al (2010) Optical bistability behaviour in a double-coupler ring resonator. In: ICAMN, international conference 2010, Prince Hotel, Kuala Lumpur, Malaysia
77. Amiri IS et al (2010) Storage of atom/molecules/photon using optical potential wells. In: International conference on experimental mechanics (ICEM) 2010, Kuala Lumpur, Malaysia
78. Amiri IS et al (2012) Multi optical Soliton generated by PANDA ring resonator for secure network communication. In: Computer and communication engineering (ICCCE) conference. IEEE Explore, Malaysia
79. Amiri IS, Afroozeh A (2014) Secure high-capacity optical communication using micro and nano-ring resonator. In: DSilva L (ed). Springer, USA
80. Amiri IS et al (2014) Single soliton bandwidth generation and manipulation by microring resonator. Life Sci J 10(12):904–910
81. Saktioto S et al (2010) Calculation and prediction of blood plasma glucose concentration. In: ICAMN, international conference 2010, Prince Hotel, Kuala Lumpur, Malaysia
82. Amiri IS et al (2014) The proposal of high capacity GHz soliton carrier signals applied for wireless commutation. Rev Theor Sci 2(4):320–333
83. Amiri IS et al (2012) Cryptography scheme of an optical switching system using pico/femto second soliton pulse. Int J Adv Eng Technol (IJAET) 5(1):176–184
84. Ali J et al (2010) Entangled photon generation and recovery via MRR. In: ICAMN, international conference 2010, Prince Hotel, Kuala Lumpur, Malaysia
85. Amiri IS, Naraei P, Ali J (2014) Review and theory of optical soliton generation used to improve the security and high capacity of MRR and NRR passive systems. J Comput Theor Nanosci (CTN) 11(9)
86. Ali J et al (2010) Dynamic silicon dioxide fiber coupling polarized by voltage breakdown. In: Nanotech Malaysia, international conference on enabling science and technology 2010, KLCC, Kuala Lumpur, Malaysia
87. Bahadoran M et al (2011) Analytical vernier effect for silicon panda ring resonator. In: National science postgraduate conference, NSPC 2011. Universiti Teknologi Malaysia, Malaysia
88. Saktioto S et al (2010) Transition of diatomic molecular oscillator process in THz region. In: International conference on experimental mechanics (ICEM) 2010, Legend Hotel, Kuala Lumpur, Malaysia

Chapter 3
Results of Digital Soliton Pulse Generation and Transmission Using Microring Resonators

3.1 Soliton Coding–Decoding by Microring Resonators Using the Optical Fiber Transmission Link

The output signal of single ring resonator for 20,000 round trips of the input signals is simulated. Signals of logic code generated from the single ring resonator can be seen from Fig. 3.1, where the optical power is fixed to 2 W at central wavelength of $\lambda_0 = 1,550$ nm and the parameters of the system are selected to $n_0 = 3.34$, $n_2 = 1.4 \times 10^{-13}$ m^2 W^{-1}, $A_{eff} = 0.25$ μm^2, $\alpha = 0.5$ dB mm^{-1}, $\gamma = 0.1$, $R = 10$ μm and $\kappa = 0.0225$. Figure 3.1a shows the output chaotic signals versus the ring round-trip, where Fig. 3.1b shows the output signals regarding to the input power. The analog and logic codes of the "0" and "1" can be generated and seen from Fig. 3.1c and d. In application, in this research the logic codes of "1010101010110101010111010111101011010101010110101" within the range of 9,050–9,100 round-trip could be generated using chaotic signals from the single ring resonator.

Thus, ring resonators are suitable to generate chaotic signals, while the logic codes (digital codes) are performed by the encrypted data. The logic codes can be transmitted over long distance communication using an encoding–decoding system. Transmitted signals can be received by the users and decoded at the end of the transmission link. The synchronously decryption of the encrypted data is processed before the chaotic codes being intercepted by the specific users via the design chaotic filters, finally, the required signals can be retrieved. The system of encoding and decoding of the transmitting signals is shown in Fig. 3.2.

Generated logic codes can be input into the encoding–decoding system. Therefore, signals in the form of encoded pulses propagate inside the optical fiber communication securely and finally can be received, detected and decoded by the users. Figure 3.3 shows the forms of transmitting signals in the communication system.

© The Author(s) 2015
I. Sadegh Amiri et al., *Soliton Coding for Secured Optical Communication Link*,
SpringerBriefs in Applied Sciences and Technology,
DOI 10.1007/978-981-287-161-9_3

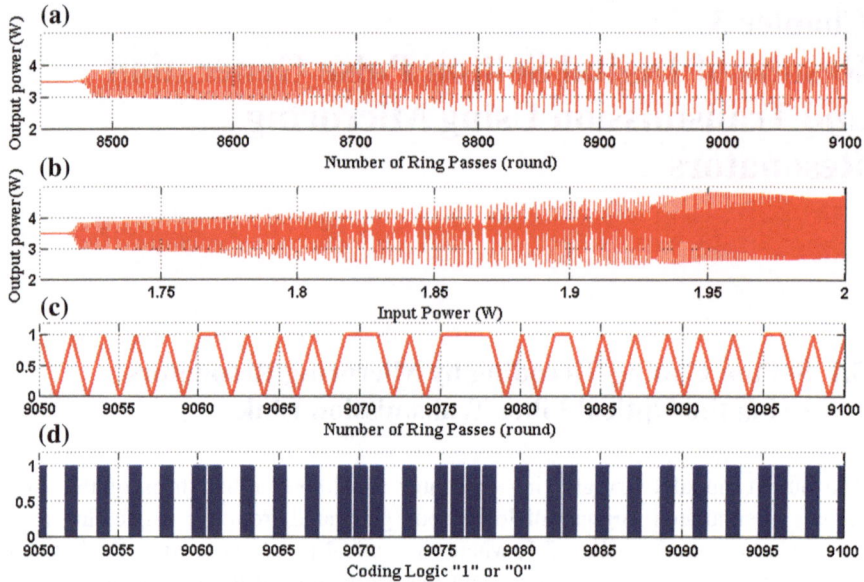

Fig. 3.1 Simulation results of chaotic signals, where **a** output power versus round-trip, **b** output power versus input power, **c** analog codes, **d** logic codes of "0" and "1"

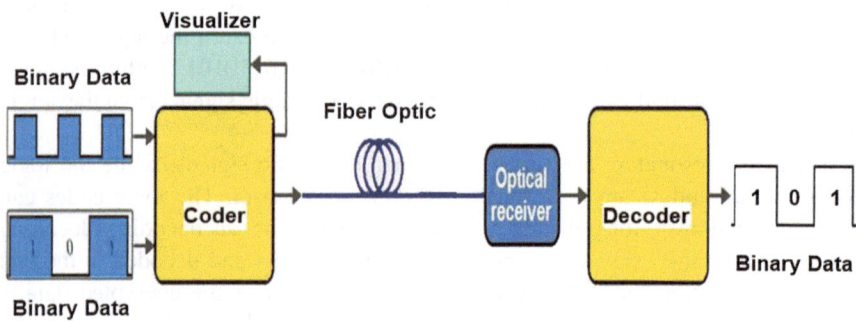

Fig. 3.2 System of encoding and decoding

The length of the optical fiber has been selected to L = 50 km, where transmitted of encoded signals can be received and detected at the end of the transmission link. Figure 3.4 shows the eye diagram of the detected signals and the decoded result. From Fig. 3.4b the original input logic codes can be retrieved which means information or data can be transmitted securely and finally be recovered with less error of the detector.

Thus, signals of chaotic can be used to generate variable codes. In this concept, we assume that the decoding of the transmitted signals can be performed by using the proposed arrangement. Optical codes via chaotic signals can be connected into

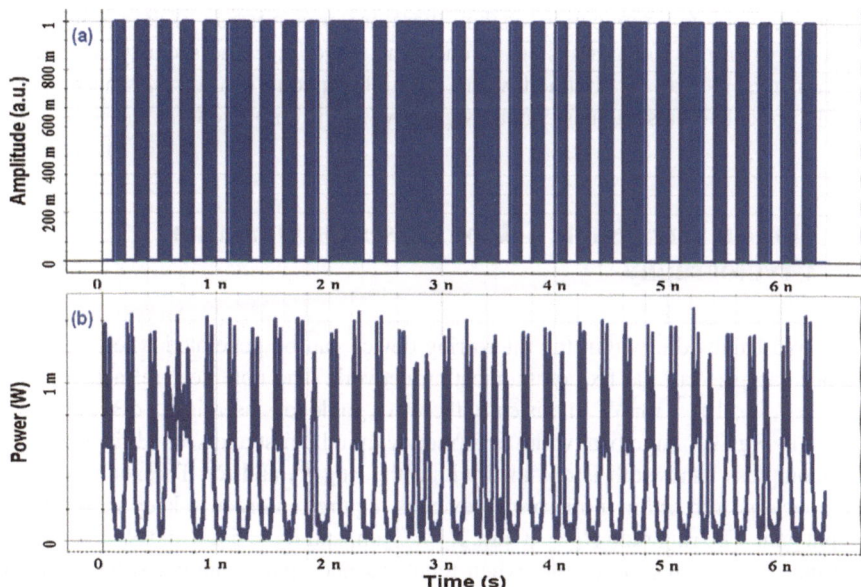

Fig. 3.3 Generation of transmitting signals where **a** signals of logic codes, **b** encoded signals

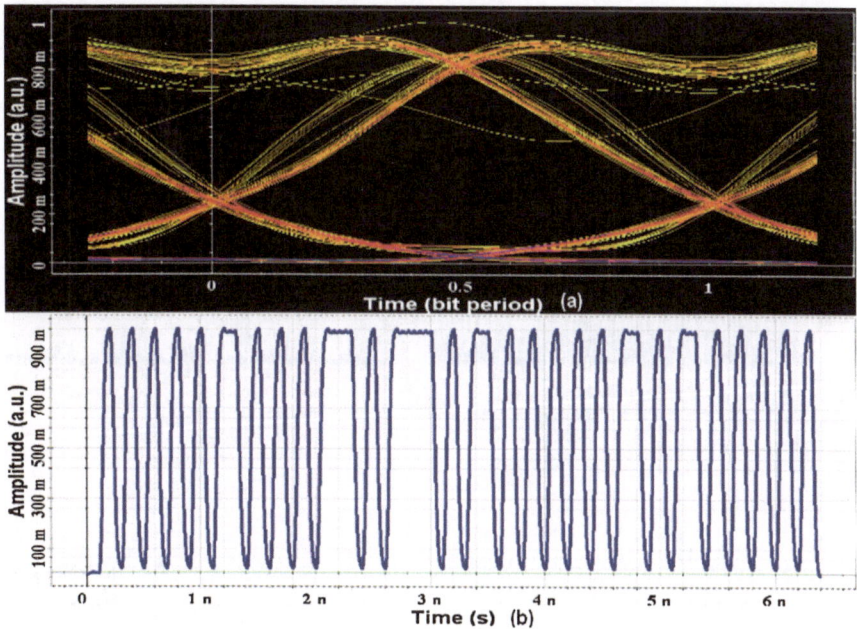

Fig. 3.4 Detection of transmitting signals where **a** eye diagram of the transmitted signals, **b** decoded signals over 50 km fiber optic communications

a fiber network communication system; therefore, transmission of data along fiber optic is performed using a system of encoding–decoding. The security scheme of the transmission can be obtained where the high capacity of transmission requires highly optical signals such as chaotic signals which is employed.

3.2 Optical Ultra-Short Soliton Pulses for Quantum Cryptography

The large bandwidth within the micro ring device can be generated where required signals can perform the fix communication network. The nonlinear refractive index is $n_2 = 1.5 \times 10^{-20}$ m²/W. In this case, the wave guide loss used is 0.5 dB mm^{-1}. As shown in Fig. 3.5, large bandwidth is formed within the first and second rings device. The compress bandwidth is obtained within the ring R_3 and R_4. The attenuation of the optical power within a microring device is required in order to keep the constant output gain. The ring radii $R_1 = 10$ μm, $R_2 = 7$ μm, and $R_3 = 5$ μm and $R_4 = 2$ μm.

Figure 3.6 shows the results when temporal and spatial optical soliton pulses are localized within a microring device and add/drop filter system with 20,000 roundtrips, where the single soliton with FWHM = 42 fs is generated. The multi soliton can be generated with FWHM and FSR of 20 Ps and 0.6 ns respectively. Here, the ring radii and their coupling coefficients are the same and $R_{ad} = 200$ μm with coupling coefficient of $\kappa_1 = \kappa_2 = 0.1$.

Thus each pair of polarization entangled photons can be made among different time frame by applying the polarization control unit and polarizer beam splitter (PBS) demonstrated in Fig. 1. They can be constituted by the two polarization

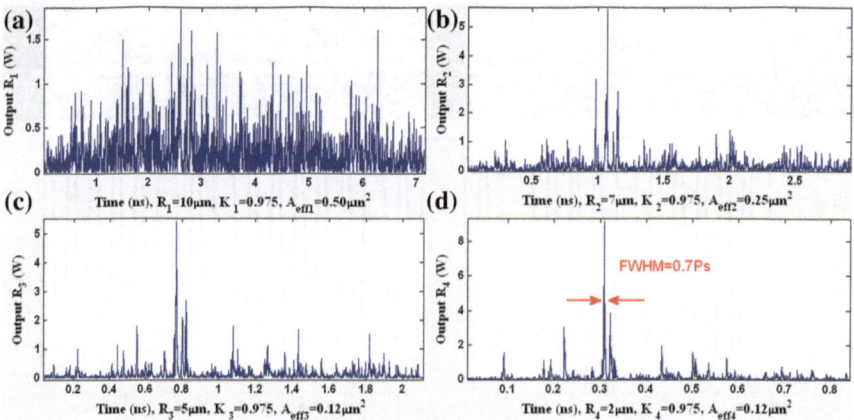

Fig. 3.5 Results obtained when temporal soliton is localized within a microring device with 20,000 roundtrips, where **a** chaotic signals from R_1, **b** chaotic signals from R_2, **c** trapping of temporal soliton, **d** localized temporal soliton with FWHM of 0.7 ps

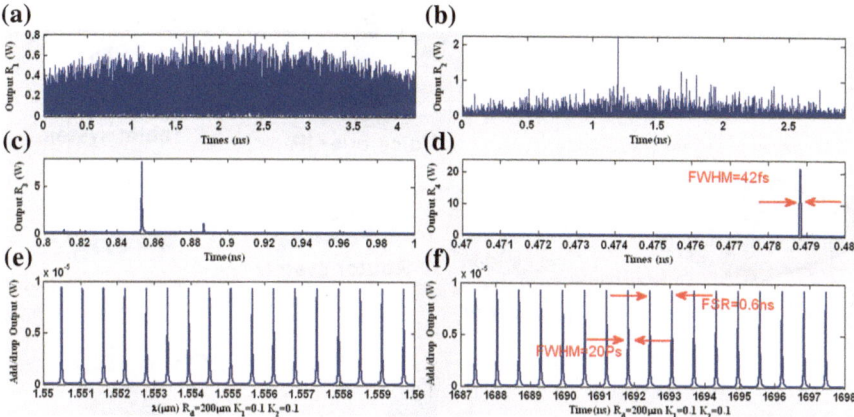

Fig. 3.6 Results of temporal and spatial soliton generation, where **a** chaotic signals from R_1, **b** chaotic signals from R_2, **c** filtering signals, **d** localized temporal soliton with FWHM of 42 fs, **e** spatial soliton pulses, **f** temporal soliton with FSR = 0.6 ns and FWHM = 20 ps

orientation angles as $[0°, 90°]$. Here we bring in the technique that is used to generate the photon. A polarization device distinguishes the basic vertical and horizontal polarization states represent to an optical switch between the short and the long pulses. We presume those horizontally polarized pulses with a temporal separation of Δt. The coherence time of the consecutive pulses is larger than Δt. Then the pursuing state at time t_1 is produced by

$$|\Phi\rangle_p = |1, H\rangle_s |1, H\rangle_i + |2, H\rangle_s |2, H\rangle_i \tag{3.1}$$

In the formula $|k, H\rangle$, k is determined as the number of time slots (1 or 2), where it announces the state of polarization (horizontal $|H\rangle$ or vertical $|V\rangle$). The subscript identifies the signal (s) or the idler (i) state. The delay circuit comprises of a coupler and the difference between the round-trip times of the Microring Resonators (MRR), which is equal to Δt. The $|H\rangle$ can be converted into $|V\rangle$ at the delay circuit output that is the delay circuits convert

$$r|k, H\rangle + t2 \exp(i\Phi)|k + 1, V\rangle + rt2 \exp(i2\Phi)|k + 2, H\rangle + r2t2 \exp(i3\Phi)|k + 3, V\rangle \tag{3.2}$$

where t and r is the amplitude transmittances to cross and bar ports in a coupler. Then equation is convinced into the polarized state by the delay circuit as

$$
\begin{aligned}
|\Phi\rangle &= \left[|1, H\rangle_s + \exp(i\Phi s)|2, V\rangle_s\right] \times \left[|1, H\rangle_i + \exp(i\Phi i)|2, V\rangle_i\right] \\
&\quad + \left[|2, H\rangle_s\right] + \exp(i\Phi s)|3, V\rangle_s \times \left[|2, H\rangle_i + \exp(i\Phi i)|2, V\rangle_i\right] \\
&= \left[|1, H\rangle_s |1, H\rangle_i + \exp(i\Phi i)|1, H\rangle_s |2, V\rangle_i\right] \\
&\quad + \exp(i\Phi s)|2, V\rangle_s |1, H\rangle_i + \exp\left[i(\Phi s + \Phi i)\right]|2, V\rangle_s |2, V\rangle_i \\
&\quad + |2, H\rangle_s |2, H\rangle_i + \exp(i\Phi i)|2, H\rangle_s |3, V\rangle_i \\
&\quad + \exp(i\Phi s)|3, V\rangle_s |2, H\rangle_i + \exp\left[i(\Phi s + \Phi i)\right]|3, V\rangle_s |3, V\rangle_i
\end{aligned}
\tag{3.3}
$$

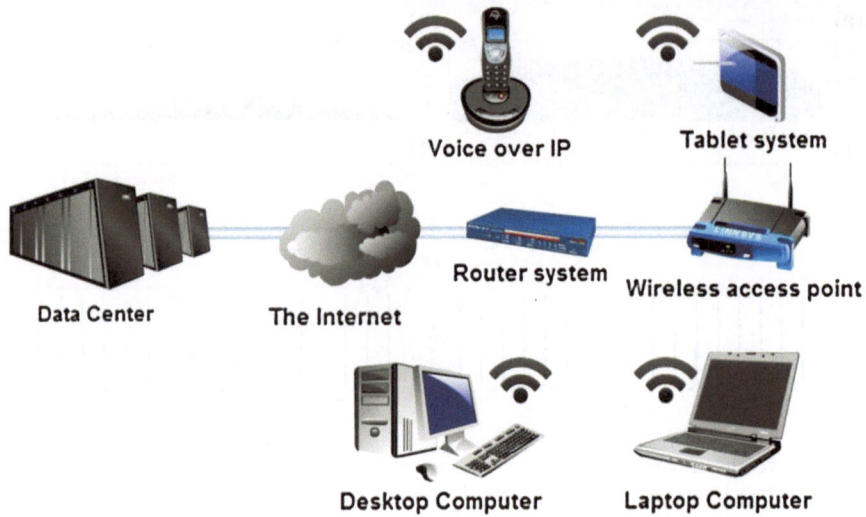

Fig. 3.7 System of optical photon transmission using a router and wireless access point

As an outcome, we can get the following polarization entangled state as

$$|\Phi\rangle = |2, H\rangle_s|2, H\rangle_i + \exp[i(\Phi s + \Phi i)]|2, V\rangle_s|2, V\rangle_i \qquad (3.4)$$

Because of the Kerr nonlinearity of the optical device, the strong pulses acquire an intensity dependent phase shift throughout propagation. The polarization angle adjustment device is utilized to investigate the orientation and optical output intensity depicted. Therefore, signal of solitons can be used to generate photon which is secured and unknown during propagation within communication systems. Generated secured photons can be transferred via a wireless access point, and network communication system shown in Fig. 3.7.

A wireless access system transmits data to different users via wireless connection. The transmission of information can be sent to the Internet using a physical, wired Ethernet connection. This method also works in reverse, when the router system used to receive information from the Internet, translating it into an analog signal and sending it to the computer's wireless adapter.

3.3 Picosecond Soliton Pulse Generation for Quantum Key Distribution Using Wired/Wireless Optical Links

A soliton pulse with a peak power of 550 mW is input into the system. The suitable ring parameters are used, for instance, ring radii $R_1 = 10$ μm, $R_2 = 10$ μm, and $R_3 = 12$ μm. The selected parameters of the system are fixed to $\lambda_0 = 87.5$ mm, $n_0 = 3.34$ (InGaAsP/InP), $A_{eff1} = 0.50$, $A_{eff2} = 0.50$ and

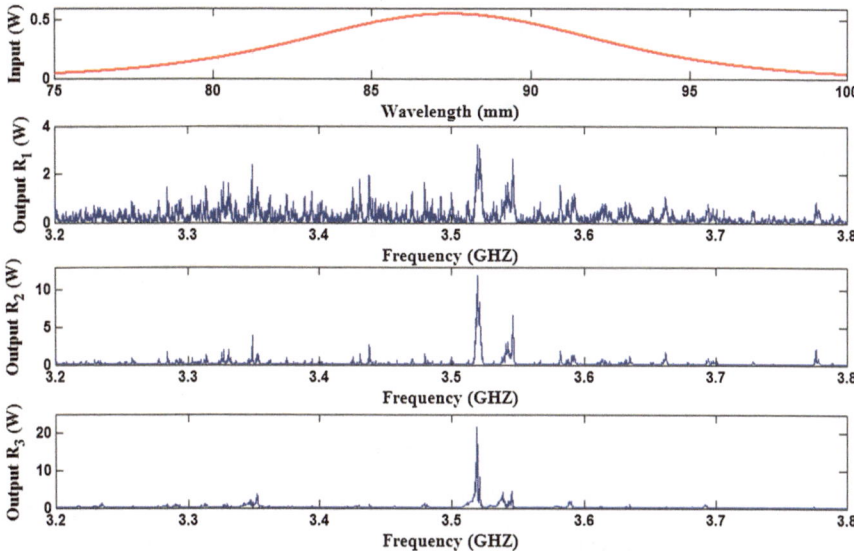

Fig. 3.8 Results obtained when frequency soliton pulse is localized within a microring device with 20,000 roundtrips, the ring radii are $R_1 = 10\ \mu m$, $R_2 = 10\ \mu m$, $R_3 = 12\ \mu m$

$A_{eff3} = 0.25\ \mu m^2$ for three MRRs respectively, $\alpha = 0.5$ dB mm^{-1}, $\gamma = 0.1$. The coupling coefficient (κ) of the MRR ranged from 0.97 to 0.998.

The large bandwidth within the microring device can be generated by using a soliton pulse input into the nonlinear MRR shown in Fig. 3.8, where the required signals perform the secure communication network. The nonlinear refractive index is $n_2 = 2.2 \times 10^{-17}$ m^2/W. In this case, the wave guided loss used is 0.5 dB mm^{-1}. From Fig. 3.8, the signal is sliced into a smaller signals spreading over the spectrum. The compress bandwidth is obtained within the ring R_2. The amplified gain is obtained within a MRR (R_3). Frequency soliton pulse is formed and trapped by using the constant gain condition. The attenuation of the optical power within a MRR is required in order to keep the constant output gain, where the next round input power is attenuated and kept the same level with the R_2 output.

Therefore ultra-short of soliton pulse can be trapped within 3.52 GHz frequency as shown in Fig. 3.9. Similarly, the temporal soliton is obtained as shown in Fig. 3.9, where a soliton pulse with peak power of 500 mW is input into the MRRs system. The ring radii $R_1 = 10\ \mu m$, $R_2 = 5\ \mu m$, and $R_3 = 4r\ \mu m$. The selected parameters of the system are fixed to $\lambda_0 = 1.55\ \mu m$, $n_0 = 3.34$ (InGaAsP/InP), $A_{eff1} = 0.50\ \mu m^2$, $A_{eff2} = 0.25\ \mu m^2$ and $A_{eff3} = 0.12\ \mu m^2$ for three MRRs respectively, $\alpha = 0.5$ dB mm^{-1}, $\gamma = 0.1$. The coupling coefficient of the MRR is fixed to 0.975. Here, a soliton pulse with FWHM of 25 ps is simulated.

Each pair of the possible polarization entangled photons is formed within different time frames by using the polarization control unit, which they can be represented by two polarization orientation angles as [0°, 90°]. They can be formed

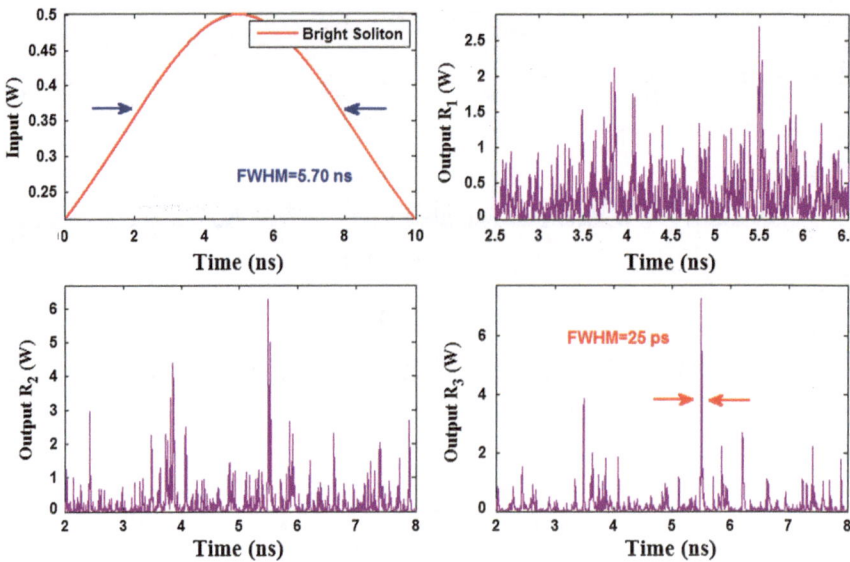

Fig. 3.9 Results obtained when temporal soliton with FWHM = 25 ps is localized within a microring device

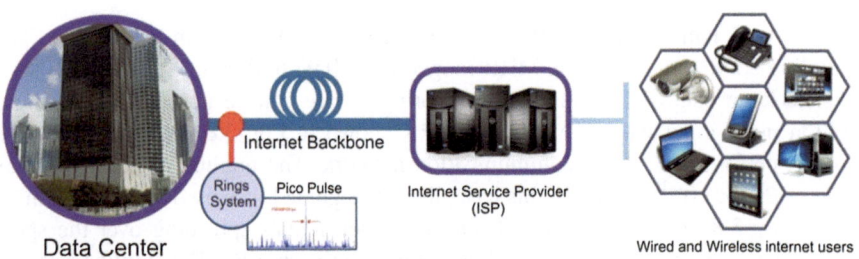

Fig. 3.10 System of entangled photons transmission using a wireless access point system

by using the optical component called the polarization rotatable device and PBS. Here we introduce the technique that is used to generate the qubits. A polarization coupler separates the basic vertical and horizontal polarization states according to an optical switch between the short and long pulses.

A wireless router system can be used to transfer generated entangled photons via a wireless access point, and network communication system shown in Fig. 3.10.

A wireless access system transmits data to different users via wireless connection. The transmission of information can be sent to the Internet using a physical, wired Ethernet connection. This method also works in reverse, when the router system used to receive information from the Internet, translating it into an analog signal and sending it to the computer's wireless adapter.

3.4 Soliton Carriers and Logic Codes for Communication Network System

The results of the chaotic signal generation are shown in Fig. 3.11. The input pulse of a soliton pulse with a power of 1 W is inserted into the system. The rings' radii and coupling coefficients are $R_1 = 5\,\mu m$, $\kappa_1 = 0.5$ and $R_2 = 3\,\mu m$, $\kappa_2 = 0.3$. The nonlinear and linear refractive indices of the system are $n_2 = 2.2 \times 10^{-17}\,m^2\,W^{-1}$ and $n_0 = 3.34$, respectively. The effective core areas of rings R_1 and R_2 are $A_{eff1} = 25\,\mu m^2$ and $A_{eff2} = 10\,\mu m^2$, respectively. Large bandwidth within the microring device can be generated by using a soliton pulse input into the nonlinear MRRs. The wave-guided loss used is $\alpha = 0.5\,dB\,mm^{-1}$, and $\gamma = 0.1$ is the fractional coupler intensity loss.

The signal is chopped (sliced) into smaller signals spreading over the spectrum; thus, a large bandwidth is formed within the first ring device. A compressed bandwidth is obtained within ring R_2. A frequency soliton pulse can be formed and trapped within the panda ring resonator with suitable ring parameters. The centred ring of the panda ring resonator system has a radius of 100 μm and coupling coefficients of $\kappa_3 = 0.35$ and $\kappa_4 = 0.30$, where the right and left rings have radii of 18 and 8 μm, respectively. The effective core areas of the right and left rings are $A_{eff1} = A_{eff2} = 0.25\,\mu m^2$, where the coupling coefficients have been selected to $\kappa_r = 0.22$ and $\kappa_l = 0.10$. Interior soliton signals inside the panda ring resonator system can be seen in Fig. 3, where the filtering and trapping processes occur during propagation of the input soliton pulse inside the system.

The chaotic soliton pulses are used widely as carrier signals in securing optical communication, wherein the information is input into the signals and ultimately can be retrieved by using suitable filtering systems. Figure 3.12a, b shows the

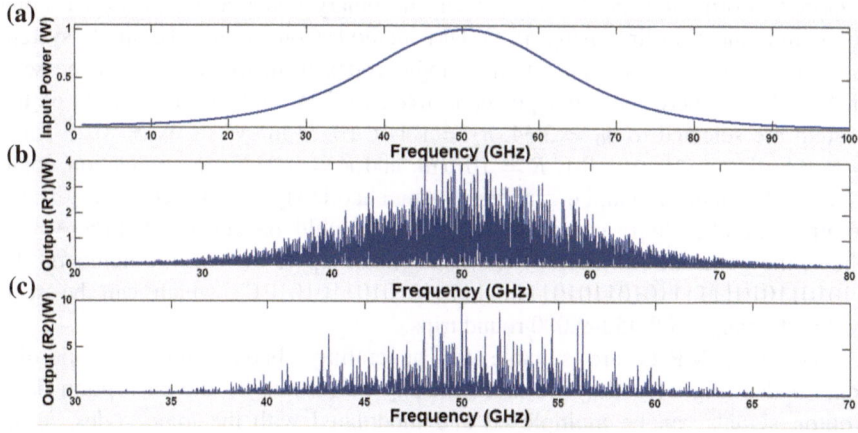

Fig. 3.11 Chaotic signal generation: **a** input power, **b** output chaotic signals from R_1, **c** output chaotic signals from R_2

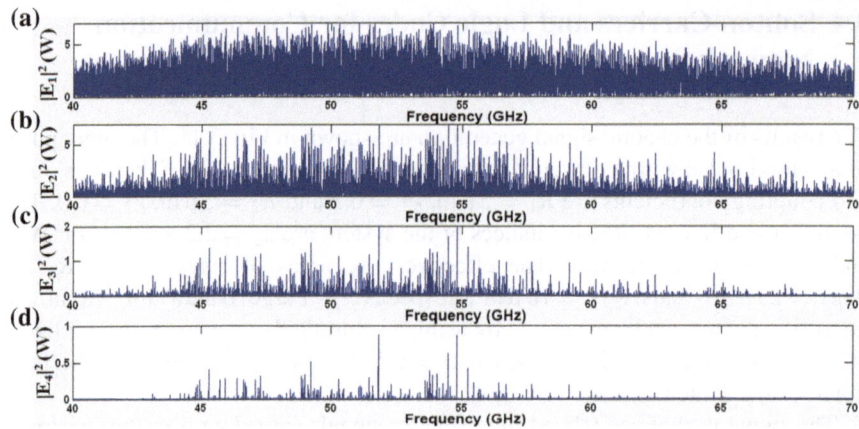

Fig. 3.12 Interior soliton powers: **a** $|E_1|^2$ (W), **b** $|E_2|^2$ (W), **c** $|E_3|^2$ (W), **d** $|E_4|^2$ (W)

interior generated signals on the right side of the panda system and Fig. 3.12c, d shows the powers of the left side.

Filtering of the interior soliton signals can be performed when the pulses pass through the couplers, κ_3 and κ_4. The output signals from the throughput and drop ports of the system can be seen in Fig. 3.13, where soliton range of 40–60 GHz are generated and used in many communication applications, such as wireless personal area networks (WPANs) and wireless local area networks (WLANs). The throughput output (E_{th}) shows localised ultra-short soliton pulses with an FWHM and FSR of 5 MHz and 2 GHz, respectively, where soliton pulses at frequencies of 50 and 52 GHz are generated. The drop output signals expressed by E_d are shown in Fig. 3.13b, c, where pulses with an FWHM of 10 MHz and FSR of 2 GHz could be generated.

A single ring resonator can be used to generate binary logic codes, where the required information can be formed via the binary codes and transmitted along the communication link using an OFDM method. Considering the single system of a ring resonator, the generation of logic codes from the system can be seen in Fig. 3.14, where the input power is fixed to 2 W and the parameters of the system are selected to $n_0 = 3.34$, $n_2 = 1.4 \times 10^{-13}$ m^2 W^{-1}, $A_{eff} = 0.25$ μm^2, $\alpha = 0.5$ dB mm^{-1}, $\gamma = 0.1$, $R = 10$ μm, and $\kappa = 0.0225$. In application, logic codes of "0" and "1" can be generated using chaotic signals. As seen in Fig. 3.14, arbitrary number logic codes of "0" and "1" could be generated. Figure 3.14 shows the generation of secret logic codes made up of the specific numbers "11 11010110101111110101101011111111010101011010101", which can be seen within the range of 8,950–9,000 round trips.

Therefore, MRRs are suitable for generating chaotic signals, while the logic codes (digital codes) are electronically performed by the encryption data. Soliton signals can be multiplexed and modulated with the logic codes, using an OFDM technique, to transmit the data via a wired/wireless network system. Transmission of data using optical soliton pulses in wireless communication has

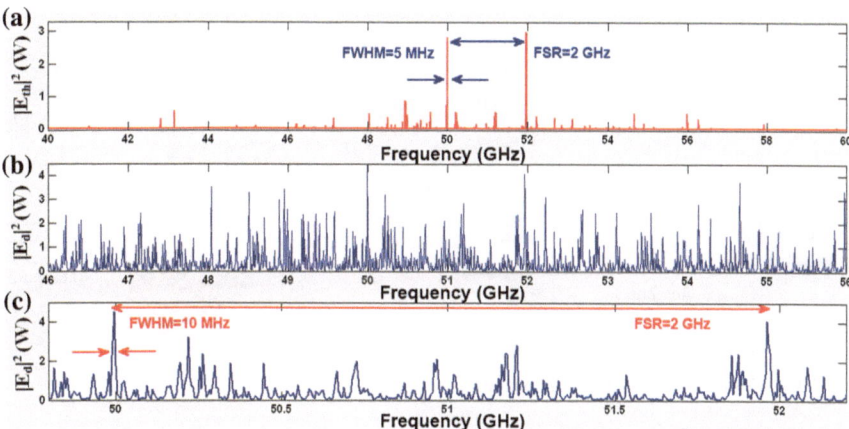

Fig. 3.13 Results of localised solitons: **a** throughput output signal with FWHM = 5 MHz and FSR = 2 GHz, **b** soliton pulses, **c** expansion of the soliton pulses ranges 50–52 GHz with FWHM = 10 MHz and FSR = 2 GHz

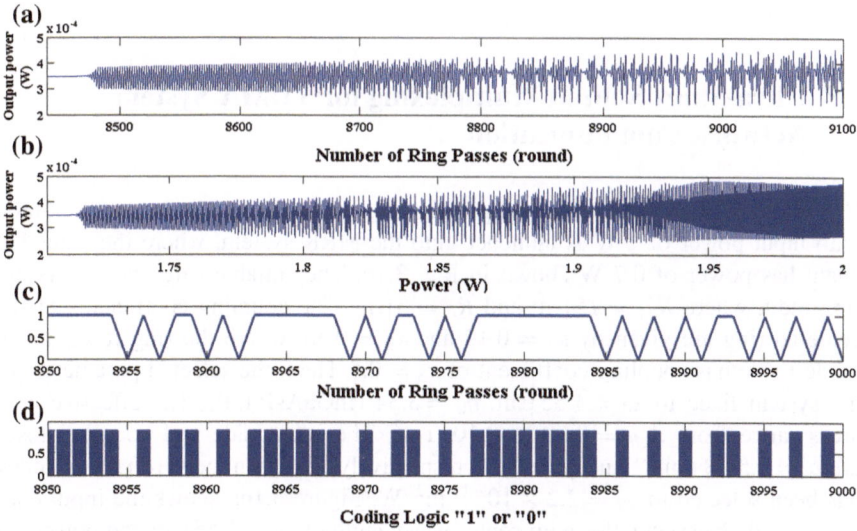

Fig. 3.14 Results of chaotic signals within the single MRR: **a** output power versus round-trip, **b** output power versus input power, **c** analog codes, **d** generated logic codes of "111101011010111 11101011010111111110101010101010101"

been investigated by several researchers. Transmitted signals can be received and de-multiplexed at the end of the transmission link. Thus, we can conduct a secure transmission of a message by logical coding of the chaotic signals, using a MRR incorporated with a communication transmission system. Such a technique can

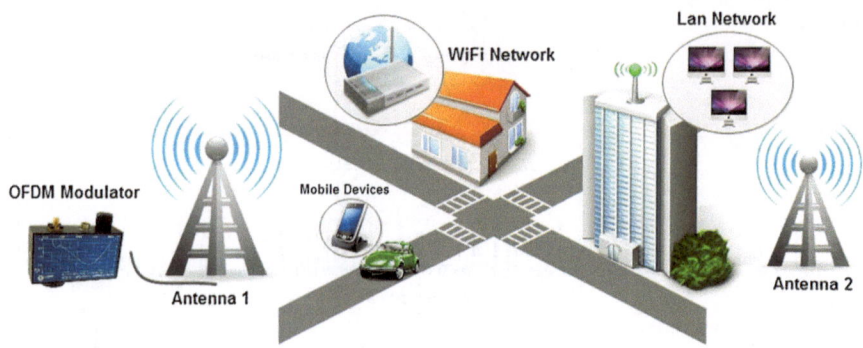

Fig. 3.15 Communication network system

overcome the problem of signal degradation, as digital signals can be recovered more easily than analog ones. The communication network system is shown in Fig. 3.15.

This high peak value causes the power amplifiers works in nonlinear part of its characteristics.

3.5 Ultra-Short Soliton Multiplexing for TDMA System Network Communication

Extremely Narrow peak of the output signals can be obtained when dark soliton with input power of 2 W is launched into the MRR system, where the Gaussian beam has power of 0.7 W shown in Fig. 3.16. The suitable ring parameters are ring radii, where $R_{ad} = 15\,\mu m$ and $R_L = 6\,\mu m$. The coupling coefficients of the centered ring are given by $\kappa_1 = 0.12$ and $\kappa_2 = 0.35$, where the ring resonator at the left side has coupling coefficient of $\kappa_3 = 0.5$. Here, the selected parameters of the system fixed to $\lambda_0 = 1.55\,\mu m$, $n_0 = 3.34$ (InGaAsP/InP). The effective core areas range from $A_{eff} = 0.50$ to $0.10\ \mu m^2$. The waveguide and coupling loses are $\alpha = 0.5\,dB\ mm^{-1}$ and $\gamma = 0.01$, respectively. The nonlinear refractive index has been selected to $n_2 = 2.2 \times 10^{-17}\ m^2/W$. Figure 3.16a shows the input dark soliton and Gaussian pulse with center wavelength of $\lambda_0 = 1.55\,\mu m$ and powers of 2 and 0.7 W respectively. Figure 3.16b, c and d show the amplified interior potential well signals, whereas the sharp pulses with FWHM of 9.57 and 8 nm can be seen in Fig. 3.16e and f for the drop and through port output signals respectively. Soliton signals can be used in optical communication where the capacity of the output signals can be improved by generation of peaks with smaller FWHM.

High capacity of transmission obtained when the optical potential wells with different center wavelengths are combined using suitable multiplexer device. Here, series of MRR systems can be integrated in one single system, incorporating with multiplexer device shown in Fig. 3.17.

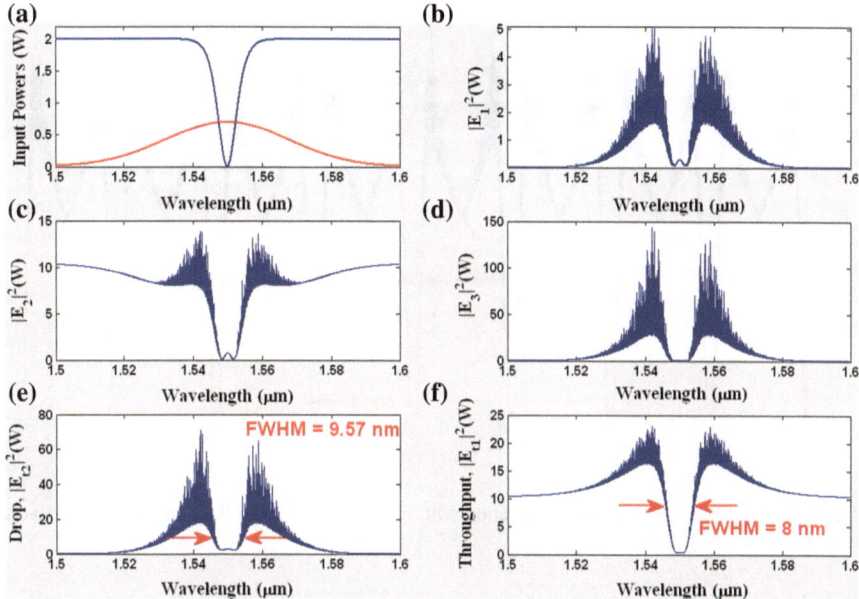

Fig. 3.16 Results of the potential wells generation **a** input dark soliton and Gaussian pulse, **b**, **c** and **d** interior amplified signals, **e** and **f** drop and through port output signals with FWHM of 9.57 and 8 nm respectively, where $\kappa_1 = 0.12$ and $\kappa_2 = 0.35$ and $\kappa_3 = 0.5$

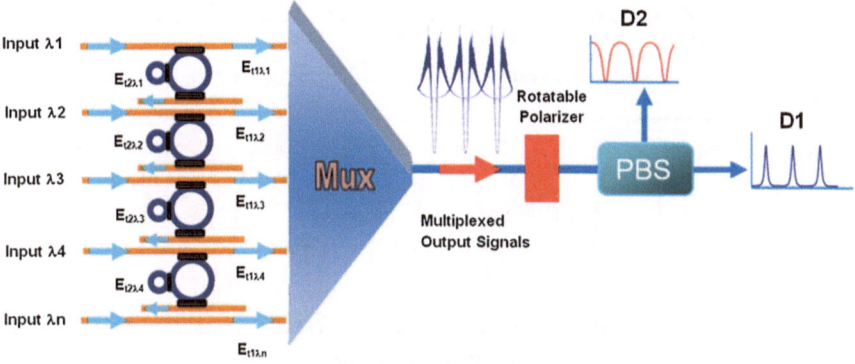

Fig. 3.17 System of integrated MRR systems, incorporating with a multiplexer device and a beam splitter

Highly potential well signals can be obtained from output of the multiplexer device, shown in Fig. 3.18. Therefore, signals with center wavelengths of $\lambda_1 = 1.53$ μm, $\lambda_2 = 1.535$ μm, $\lambda_3 = 1.54$ μm, $\lambda_4 = 1.545$ μm, $\lambda_5 = 1.55$ μm, $\lambda_6 = 1.555$ μm, $\lambda_7 = 1.56$ μm, $\lambda_8 = 1.565$ μm and $\lambda_9 = 1.57$ μm are combined, where pulses with FWHM and FSR of 0.8 and 5 nm can be obtained respectively.

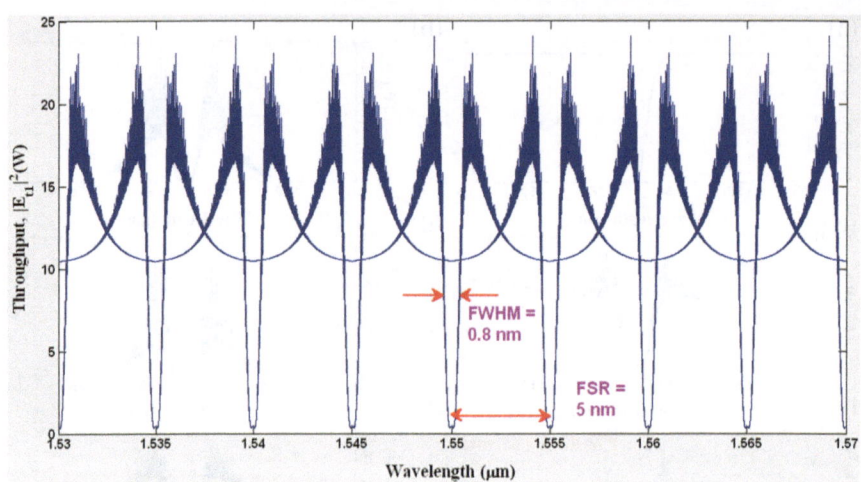

Fig. 3.18 Multiple potential well generation with FWHM and FSR of 0.8 and 5 nm respectively, using multiplexer system

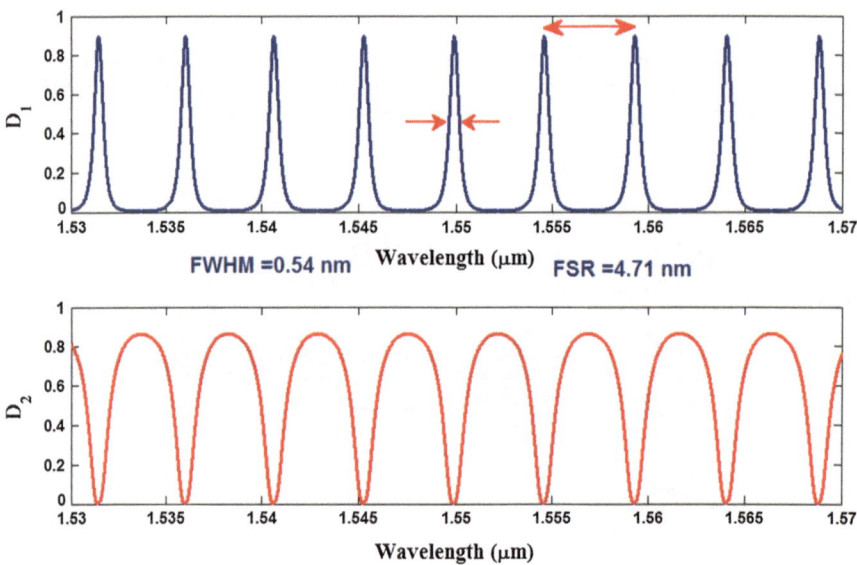

Fig. 3.19 Dark and bright soliton generation with FWHM and FSR of 0.54 and 4.71 nm respectively, using multiplexer system

In order to generate quantum binary and logic codes of "0" and "1", the multiplexed signals from the multiplexer system transmit into a beam splitter (PBS). In operation, the dark and bright soliton can be generated within the proposed system after traveling of the signals through the PBS shown in Fig. 3.19, whereas the polarization phase shift of the two components is 90°.

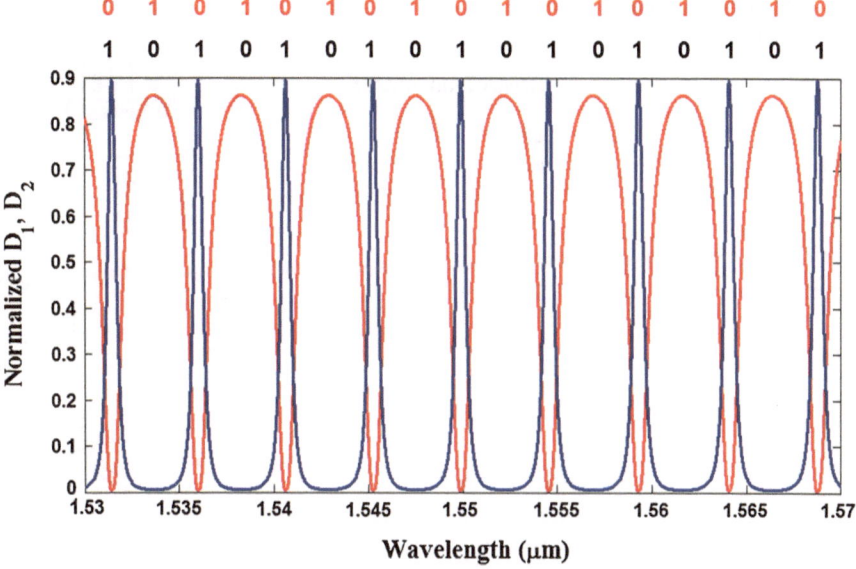

Fig. 3.20 Dark and bright soliton pulses simultaneously seen from photo detectors 1 and 2

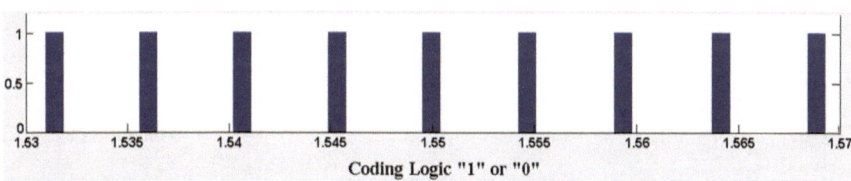

Fig. 3.21 Generation of logic codes, using PBS

Figure 3.20 show the normalized binary dark and bright soliton pulses which seen simultaneously from photo detectors 1 and 2. These optical pulses can be converted to digital binary codes of "0" and "1" using suitable analog to digital convertor electronic device.

Therefore, random polarization states of two components can be used to form the binary code patterns and the binary code signals. The logic states are set as shown in Fig. 3.21.

Here, the logic codes detected by D_1 are '10101010101010101' patterns. The logic codes detected by D_2 are '01010101010101010' patterns. Different orders of the logic codes can be used to generate different signal information, propagating in the network communication via TDMA transmission system shown in Fig. 3.22. This system uses data in the form of secured logic codes to be transferred into the singular users via different length of the fiber optics line to the TDMA transmitter.

Therefore, same digital information of codes can be shared between users with different time slots. The transmission unit is a part of quantum processing

Fig. 3.22 Schematic of the
TDMA system

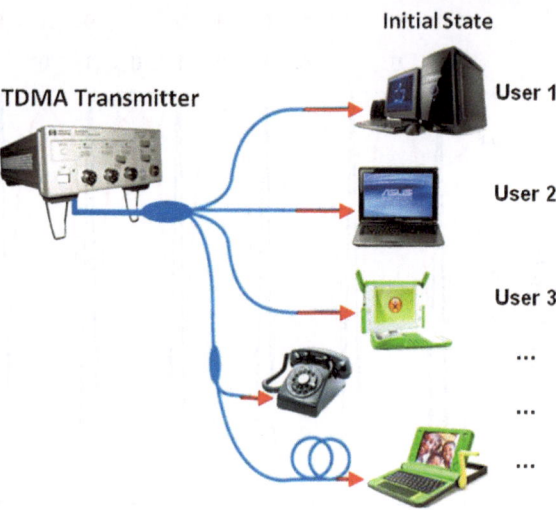

system that can be used to transfer high capacity packet of quantum codes using
extremely short pulses of dark and bright soliton. Moreover, the high capacity of
data can be applied by using more wavelength carriers.

Conclusion

In conclusion, we have presented nonlinear effects of a single ring resonator known as optical chaos. Gaussian laser pulse with central wavelength of 1.55 μm is inserted into the system generating high capacity of chaotic signals. The optical input power was fixed to 2 W, where the parameters of the system have been selected to $n_0 = 3.34$, $n_2 = 1.4 \times 10^{-13}$ m^2 W^{-1}, $A_{eff} = 0.25$ μm^2, $\alpha = 0.5$ dB mm^{-1}, $\gamma = 0.1$, $R = 10$ μm and $\kappa = 0.0225$. Transmission of signals can be implemented via an encoding-decoding method where the encoded signals of the logic codes can be obtained and secured during transmission along long distance fiber optic communication. Here the logic codes of "101010101011010101011101011110 10110101010101010110101" could be generated from chaotic signals using a ring resonator system. The decoding of signals can be obtained at the end of the transmission link. Here the length of the transmission link has been selected for 50 km where clear and decoded signals were achieved by the users thus providing secured and high capacity of optical soliton communication. Photon can be performed using single and multiple temporal and spatial soliton pulses generated by MRR system. We have shown that a large bandwidth of the arbitrary soliton pulses can be generated and compressed within a micro waveguide. The chaotic signal generation using a soliton pulse in the nonlinear MRRs has been presented. Localized light of soliton perform secure and high capacity of optical communication. Localized spatial and temporal soliton pulse is useful to generate entangled photon pair providing quantum key applicable for communication wireless networks. In this study ultrashort of single optical soliton with FWHM = 0.7 ps and 42 fs and multi optical soliton which FWHM and FSR of 20 ps and 0.6 ns were generated propagating along the entangled photon generation system which is connected to the drop port of the add/drop filter system connecting to series of MRRs.

Thus we have analyzed the entangled photon generated by chaotic signals in the series MRR devices applicable in optical wireless communication systems. An interesting concept can be presented, in which the quantum key is generated by using a remarkably simple system. Proposed system consists of a series of MRRs.

© The Author(s) 2015
I. Sadegh Amiri et al., *Soliton Coding for Secured Optical Communication Link*,
SpringerBriefs in Applied Sciences and Technology,
DOI 10.1007/978-981-287-161-9

Balance between dispersion and nonlinear lengths of the soliton pulse exhibits the self-phase modulation effect. Therefore, light pulse can be trapped, localized coherently within the waveguide. We have shown that a large bandwidth of the arbitrary soliton pulses can be generated and compressed within a micro wave-guide. The chaotic signal generation using a soliton pulse propagating within the nonlinear MRR has been presented. A selected light pulse can be localized and used to perform the secure communication network. Localized spatial and temporal soliton pulse can used to generate entangled photon pair that provides quantum keys, applicable for communication networks. We have analyzed the entangled photon generated by chaotic signals in the series MRR devices. The classical information and security code can be formed by using the temporal and spatial soliton pulses, respectively, where the transmission of secured data can be implemented via wireless access point system used in optical communication networks. An MRR system incorporating a panda ring resonator has been demonstrated. An optical soliton is generated by the bright soliton pulse propagating within the presented system.

A high frequency band of optical soliton pulses can be generated and used in optical communication networks, where soliton carriers are performed in order to be modulated and multiplexed with logic data codes generated by a single ring resonator. Thus, high bit rate data transmission can be provided, using a broad soliton frequency band. The OFDM technique is used to transmit the data of logic codes via a wired/wireless communication network. The efficiency of the system was evaluated using the CCDF measurement curve. Also in order to investigate the guard interval for combating inter-symbol interference in OFDM systems, two scenarios based on FFT and DWT for implementing the system were discussed which shows the better performance of DWT-OFDM system.

We have analyzed and described dark-Gaussian soliton collision behaviors within a modified add/drop multiplexer system consisting of one center ring and one smaller ring on the left side. Optical tweezer in the form of potential well could be generated and used to perform binary codes using PBS. We have proposed an interesting concept of internet security based on quantum logic codes where the use of data encoding for high capacity communication via optical network link is plausible. Required binary codes could be generated after the potential well signals were travelling into the PBS. The potential wells are highly secured optical signals because of low intensity of their center wavelengths in which detection of such as signals is extremely difficult. Therefore, the importance of the study is that the generated potential wells signals are highly secured along the MRRs system, where a part of the system consists of transmission part can be used to transfer of the quantum codes. Results obtained have shown that the multiplexed signals of potential wells with FWHM and FSR of 0.8 and 5 nm can be performed and used to logic codes generation, where the dark and bright soliton pulses with FHHM and FSR of 0.54 and 4.71 nm could be obtained. Furthermore, such a concept is also available for hybrid communications, for instance, wire/wireless and satellite.